METAL BUTTONS

c.900 BC – c.AD 1700

METAL BUTTONS

c.900 BC – c.AD 1700

by

Brian Read

principal illustrators

Patrick Read and Nick Griffiths

Monumental brass of Thomas and Edith Polton, AD 1418,
Wanborough Church, Wiltshire. After Kite 1860

Portcullis Publishing

First published 2005 by
Portcullis Publishing, Meadow View, Wagg Drove, Huish Episcopi, Langport,
Somerset, TA10 9ER, United Kingdom.

This corrected, revised and enlarged second edition published 2010 by
Portcullis Publishing, Meadow View, Wagg Drove, Huish Episcopi, Langport,
Somerset, TA10 9ER, United Kingdom.

ISBN 9780953245062

Book and cover design by Philippa Foster, 5D Illustration,
Somerset, United Kingdom.
Typeset and printed by Short Run Press Limited, Exeter, Devon, United Kingdom.

By the same author:
History Beneath Our Feet, 1988. Merlin Books.
History Beneath Our Feet, 1995. Anglia Publishing.
Cockington Bygones Vol 1, 1999. Portcullis Publishing.
Cockington Bygones Vol 2, 2000. Portcullis Publishing.
Metal Artefacts of Antiquity Vol 1, 2001. Portcullis Publishing.
Cockington Bygones Vol 3, 2003. Portcullis Publishing.
Metal Buttons c.900 BC – c.AD 1700, 2005. Portcullis Publishing.
Hooked-Clasps and Eyes, 2008. Portcullis Publishing.

Brian Read was born in 1939 in Essex and raised in East and South-East London. With no formal educational qualifications, in 1954 he left Secondary Modern school and became a trainee millwright and then a trainee groundsman before joining the Merchant Navy in 1955 where he travelled widely. In 1961 he embarked on a fire service career, first with the Devon County Fire Service, then the City of Plymouth Fire Brigade, and finally the newly formed Devon Fire Brigade. While on duty in 1983, in the rank of assistant divisional officer, he sustained an injury that, in 1986, resulted in his medical discharge.

Since leaving the fire service he has worked as a freelance writer. His first book *History Beneath Our Feet*, published in 1988, was a best seller and after extensive revision underwent re-publication in 1995 and again proved successful. In 1989 he extended into self-publishing, using the imprint Portcullis Publishing: this second edition of *Metal Buttons c.900 BC – c.AD 1700* is the seventh title under this imprint.

For more than three decades, metal-detecting and its associated study of small metal artefacture has been his primary leisure interest.

Knowledge unshared is knowledge lost

Dedication

To Val, for her support and encouragement

Contents

	Page:
Notes on Illustrations	xiii
Drawing Conventions and Abbreviations	xiv
Acknowledgements	xv
Foreword	xix
Introduction	xxi

1: Late Bronze Age Buttons — 1
Cast one-piece copper-alloy button-like objects with integral undrilled shanks.
Cast one-piece copper-alloy button-like object with integral drilled shank.
Probable composite two-piece die-stamped or repoussé sheet copper-alloy
and shale button-like object with indeterminate shank.
Composite three-piece sheet gold and shale button-like object with (?) stitching-holes.

2: Early Iron Age Buttons — 4
Composite two-piece die-stamped or hammered sheet copper-alloy button-like
objects with probable separate undrilled shanks.

3: Late Iron Age and Roman Period Toggle-Fasteners — 5
Cast one-piece copper-alloy toggle-fasteners with integral undrilled shanks
or shank-less.

4: Late Iron Age and Roman Period Button-and-Loop Fasteners — 7
Cast one-piece copper-alloy button-and-loop fasteners with integral
undrilled shanks.

5: Roman Period Buttons — 11
Cast one-piece copper-alloy button-like object with integral drilled shank.
Cast one-piece copper-alloy button-like object with possible integral drilled shank.
Cast one-piece copper-alloy button-like object with integral undrilled shank.

6: Medieval Buttons — 13
Cast one-piece copper-alloy buttons with integral drilled shanks.
Cast one-piece copper-alloy buttons, with integral undrilled shanks.
Cast one-piece lead/tin alloy or tin buttons with integral undrilled shanks.
Composite two-piece cast copper-alloy buttons with separate embedded drawn
copper-alloy wire shanks.
Composite two-piece cast lead/tin alloy buttons with separate embedded
drawn copper-alloy wire shanks.

Composite three- or five-piece die-stamped sheet copper-alloy buttons with
separate soldered drawn copper-alloy wire shanks.

Composite three-piece die-stamped sheet copper-alloy button with separate soldered
sheet copper-alloy shank.

Composite three-piece die-stamped sheet copper-alloy buttons with separate
soldered drawn copper-alloy wire shanks.

Composite three-piece die-stamped sheet copper-alloy and sheet iron button with
separate soldered drawn copper-alloy wire shank.

Composite three-piece die-stamped sheet silver button with separate soldered drawn
silver wire shank.

Composite three-piece die-stamped sheet copper-alloy buttons with lead/tin
alloy-filled backs and separate embedded drawn copper-alloy wire shanks.

7: Post-Medieval Buttons 35

Cast one-piece copper-alloy buttons with integral drilled shanks.

Cast one-piece copper-alloy buttons or cloak-fasteners with integral drilled shanks.

Cast one-piece silver button or cloak-fastener with integral drilled shank.

Cast one-piece copper-alloy buttons with integral undrilled shanks.

Cast one-piece lead/tin alloy buttons with integral drilled shanks.

Cast one-piece lead/tin alloy buttons with integral undrilled shanks.

Cast one-piece silver button with integral undrilled shank.

Composite two-piece cast copper-alloy buttons with integral drilled shanks.

Composite two-piece cast copper-alloy button or cloak-fastener with integral
drilled shank.

Composite two-piece cast copper-alloy and sheet iron button with integral
drilled shank.

Composite two-piece cast copper-alloy and lead/tin alloy button or cloak-fastener
with integral undrilled shank.

Cast composite two-piece copper-alloy buttons with separate embedded drawn
copper-alloy wire shanks.

Composite two-piece cast copper-alloy buttons with separate embedded drawn iron
wire shanks.

Composite two-piece cast lead/tin alloy buttons with separate embedded drawn
copper-alloy wire shanks.

Composite two-piece cast lead/tin alloy buttons with separate embedded drawn
iron wire shanks.

Composite two-piece cast copper-alloy buttons or cloak-fasteners with separate
soldered drawn iron wire shanks.

Composite two- or three-piece lead/tin alloy buttons with drawn iron or
indeterminate shanks.

Composite three-piece die-stamped sheet copper-alloy buttons with separate
soldered drawn copper-alloy wire shanks.

Composite four-piece repoussé sheet silver button with separate soldered drawn silver wire shank.

Composite three-piece die-stamped sheet silver buttons with separate soldered drawn silver wire shanks.

Composite two-piece cast silver button with separate embedded drawn silver wire shank.

Composite silver openwork filigree buttons with separate soldered drawn silver wire shanks.

Composite three-piece silver infilled openwork filigree and sheet copper-alloy button with separate soldered drawn copper-alloy wire shank.

Composite two-piece cast or sheet silver-gilt ground-supported filigree hanging-buttons with separate soldered drawn silver wire shanks.

Composite drawn silver wire button with separate drawn silver wire indeterminate double shank.

Composite three-piece part cast part die-stamped copper-alloy buttons with separate soldered drawn copper-alloy wire shanks.

Composite three-piece cast copper-alloy and sheet iron button with separate soldered drawn copper-alloy wire shanks.

Composite four-piece drawn wire or sheet copper-alloy buttons with separate soldered drawn copper-alloy wire or sheet copper-alloy shanks.

8: Post-Medieval Cuff Links 128

Cast one-piece silver cuff-link with integral drilled shank.

Cast one-piece copper-alloy cuff-link with integral drilled shanks.

Composite three-piece sheet copper-alloy cuff-link with separate drawn wire shank.

Cast one-piece lead/tin alloy cuff-links with integral undrilled shanks.

Bibliography 131

Other Illustrations

Front cover

Seventeenth-century buttons, and one of the angels painted on the ceiling of
Muchelney Church, Somerset.

Back cover

Medieval buttons.

Frontispiece

A cornucopia of early metal buttons.

Half-title

Monumental brass of Thomas and Edith Polton, AD 1418,
Wanborough Church, Wiltshire.

Fig 1. The Camomile Street Soldier, detail; in the MoL.

Fig 2. Monumental brass of Robert Wyvil, Bishop of Salisbury (died 1375), in Salisbury
Cathedral. The Bishop's champion is depicted standing outside the door of
Sherborne Castle, Dorset.

Fig 3. Monumental effigy of perhaps Sir Richard de Willoughby (Edward III period),
in Willoughby Church, Nottinghamshire.

Fig 4. A monumental brass of Sir Miles (died 1344) and Lady Joan de Stapleton,
formerly in Ingham Church, Norfolk.

Fig 5. Monumental brass of Sir John de Northwood (died 1337), and his wife
Joan de Badlesmere, in Minster Church, Isle of Sheppey, Kent.

Fig 6. Monumental effigy of Joan Burwaschs, Lady Mohun, 14th century, in the
Lady Chapel, the undercroft of Canterbury Cathedral, Kent.

Fig 7. Monumental effigy of William of Windsor and Blanche de la Tour, children of
Edward III, c.14th century, in the Chapel of St Edmund, Westminster Abbey,
London.

Fig 8. Figures on the tomb of Sir Roger de Kerdeston (died 1337), representing relatives
of the deceased, in Reepham Church, Norfolk.

Fig 9. Robert Dudley, Earl of Leicester 1532?-88, anonymous artist, English, *c.*1570, in Montacute House, Somerset, NT.

Fig 10. Sir Thomas Cavandish 1560-92, buccaneer and the second Englishman to circumnavigate the world (1586-88), English, late 16th century, in Montacute House, Somerset, NT.

Fig 11. The almost unique painted ceiling of Muchelney Church, Somerset; depicting ten angels, early 17th century.

Fig 12. Viscount Cobham, 1730, redrawing of contemporary engraving.

Fig 13. A stumpwork mirror depicting Charles II and Catherine of Braganza, *c.*1670, in SSWM.

Fig 14. Part of the stumpwork mirror detail enlarged.

Fig 15. Mercers' Maiden from the Warden's Book of Arms 1634.

Fig 16. Contemporary painting of the Duke of Monmouth, oil on canvas, in SCMT.

Fig 17. Boxed shoe-buckle and silver button reputedly worn by the Duke of Monmouth, in SCMT.

Notes on Illustrations

Each button, button-like object or other fastener is illustrated either by a line-drawing, a photograph, or a line-drawing and a photograph. With several exceptions, which are either enlarged or slightly smaller, line-drawings are approximately actual-size, while photographs are not to scale. Some photographs are much enlarged to show detail. Where known, metric approximate dimensions are included within the text. Every artefact has an individual catalogue number. Museum accession numbers, Portable Antiquities Scheme identification and United Kingdom Detecting Finds' Database numbers, copyright acknowledgements and relevant bibliographical references are mentioned individually. Some objects which have distorted sections, for example shanks, are drawn as they were originally intended.

Line-drawing no. 1 reproduced courtesy of Paul Robinson. Line-drawings nos 2, 9-11, 13, 18-21, 24, 35-37, 40-45, 49, 53, 55, 60, 63, 78-80, 89-90, 96, 99, 111-13, 128-31, 133-44, 147-49, 153-79, 181, 184, 189, 192-95, 197-99, 201-02, 205-13, 222-29, 235-41, 244-56, 262-79, 281, 283-85, 287, 290-98, 306-10, 312-13, 315-27, 329, 332-63, 365-93, 395-97, 399-401, 403, 412-17, 420, 423, 425-30, 434-35, 443-46, 448-52, 456-58, 460, 462-68, 477-78, 480-81, 490, 492, 498, 500-04, 507 and photograph no. 194 © Patrick Read; figs nos 1, 12 and line-drawings nos 12, 16, 27, 47-48, 54, 57, 68, 76, 81-83, 107, 145, 150-52, 180, 182, 186, 188, 190-91, 196, 203, 242-43, 282, 304-05, 394, 411, 421, 424, 442, 455, 479, 484-87, 505 © Nick Griffiths; line-drawing no. 328 © Anne Hodgson; line-drawings nos 151-52, 180, 182, 190-91 reproduced courtesy of Tony Pilson; photographs no. 327 © and reproduced courtesy of Ron Morley; photograph no. 364 © and reproduced courtesy of Ian Chubbock; line-drawing no. 26 © and reproduced courtesy of Kent County Council (illustrator, Dominic Andrews); photographs nos 7-8 © and reproduced courtesy of the British Museum; line-drawings nos 52, 69-75, 91, 93-95, 97 © and reproduced courtesy of the Museum of London; photograph and drawing of no. 183 © of the Portable Antiquities Scheme and Lancashire County Council; figs 13-14 © and reproduced courtesy of Salisbury and South Wiltshire Museum; photograph no. 489 and figs 16-17 reproduced courtesy of Somerset County Museums Service; photograph no. 194 reproduced courtesy of Torquay Museum; fig. 15 © and reproduced courtesy of the Mercers' Company; photograph no. 25 © and reproduced courtesy of English Heritage, Hadrian's Wall Museums, Chesters; figs 9-10 reproduced courtesy of the National Trust, Montacute House, Somerset; all other photographs and figures © Brian Read. Reproduction of these illustrations is prohibited without express permission in writing from the respective copyright owner.

Drawing conventions

Standard drawing conventions describe enamel colours, however, where the small size precludes their use, the text includes a description.

Red	Blue	Blue	Blue	White

Abbreviations used within the text

D	diameter
L	length
W	width
acc. no.	accession number
PAS	Portable Antiquities Scheme
UKDFD	United Kingdom Detecting Finds' Database
pers. comm.	personal comment
BM	British Museum
V&A	Victoria and Albert Museum
MoL	Museum of London
WHM	Wiltshire Heritage Museum
WANHS	Wiltshire Archaeological and Natural History Society
SSWM	Salisbury and South Wiltshire Museum
SCMT	Somerset County Museum, Taunton
NT	National Trust
DA	*Dress Accessories c.1150 – c.1450 Medieval Finds From Excavations In London: 3. 1991*
FFRA	Finds from Roman Aldborough. A Catalogue of Small Finds from the Romano-British Town of Isurium Brigantum. *Oxbow Monograph 65. 1996*

Acknowledgements

The popularity of *Metal Buttons c.900 BC – c. AD 1700*, coupled with an accumulation of fresh material evidence, prompted the compilation of this second edition. Again, it would have been impossible without the enthusiastic co-operation of many people and it is to them that the real credit is due, especially the many private landowners throughout the country who permit responsible metal-detecting on their respective properties: but for them our knowledge of small metalwork would be the poorer.

For allowing their buttons and button-like objects to be recorded for inclusion herein, or bringing to my attention an interesting item, my gratitude is extended to the following independent metal-detectorists and members of metal-detecting clubs: Richard Berry, Rod Blunt, David Button, Steve Button, Peter Chamberlain, Ian Chubbock, Tim Cooke, Paul Cowburn, Garry Croucher, Gillian Davies, Allan Emmerson, Tony Gibson, Ron Heaps, Richard Humphrey, the McGeachy family, Ray Merrell, Anthony Mims, C Openshaw, Chris Osborne, Mike Pegg, Jon Petter, Greg Ramsey, Colin Richardson, Dave Shelley, Margaret Sherlock, Barry Sherlock and Patrick Thorn: Border Reivers Search Society – Alan Donaldson; Canterbury and District Searchers – Anthony Lowe; Cotswold Heritage and Detecting Society – John Bromley; East Devon Metal-Detecting Club – Jim Cobley, Ron Gibson, Gordon Glover, Mike Green, Colin Hart, Dave Hewett, Mark Hanley, Dave Kerr, Robert Killick, Paul Maeer, Pauline Miles, Ian McFadzean, Chris Mogford, Doug Oseland, Alan Stevenson, Nigel Tucker, Paul Urqhart, Steve Waterall, Roger Weaver, Sam Weller, Paul Wood and John Wright; Fen and Wolds Metal-Detecting Club – Andy Germaney; Fenland Finders Metal-Detecting Club – Cheryl Hodgson; Hinkley Search Society – John Caluori and Tessa Caluori; Lincoln Historical Search Society – Rob Lane; Mid-Kent Metal-Detecting Club – Nick Hampshire; Norwich Detectors – Ron Morley; Quakers Acres Metal-Detecting Club – Terry Wayne; Society of Thames Mudlarks – Ken Bellringer, Chaz Bullock, Vic Housten, John Mills, Stephanie Mills and Tony Pilson; Stour Valley Search and Recovery Club – Andy Mitchell, Bob Tydeman and Ken Wheatley; Torbay Metal-Detecting Club – Fred Brown; Weymouth and Portland Metal-Detecting Club – Dave Cobb, Alan Davies, Simon Gover, Bob Needham, Paul Rainford, Martin Savage, Paul Shannon and Steve Wootton; Yeovil Metal-Detecting Club – Chris Adams, Ken Bellringer, Anne Laverty, Mark Cowan, Roger Evans, Graham Libbey, Robert Lovett, Val MacRae, Bob March, Alan Maidment, Derek Malkin, Lucy Miller, Anne Morgan, Mike Otterbeck, Maxine Palfreeman, Mark Pickersgill, Mike Pittard, Paul Rainford, Alan Riste, Paul Shukri, Gordon Sinfield, Andrea Thompson, Hugh Vincent and Andy Wass.

The curatorial and other staff of the following museums and institutions provided photographs and line-drawings, allowed the viewing and photographing of buttons in their respective collections, or provided invaluable opinions or other assistance, for which I am grateful: Bath Museum of Costume – Elaine Uttley; British Museum – Peter Higgs, Alan Scollan and museum assistants (Department of Greek and Roman

Antiquities), Dr Ralph Jackson (Department of Prehistory and Europe) and Kellie Leydon (Photography and Imaging); English Heritage, Hadrian's Wall Museums, Chesters – Georgina Plowright; Liverpool Museums – Janny Baxter; National Army Museum – Craig Murray; Museum of London – Geoff Egan (Specialist Services) and Catharine Maloney (Archaeological Archive); National Trust, the National Portrait Gallery collection at Montacute House, Somerset – Graham Meadon; the Wade Costume Collection, Berrington Hall, Herefordshire – Althea Mackenzie; and the Gallery of Costume, Platt Hall, Manchester – Dr Miles Lambert; Portable Antiquities Scheme – Dot Boughton, Stuart Noon, Andrew Richardson, Ciorstaidh Hayward Trevarthen and David Williams; Salisbury and South Wiltshire Museum – Martin Wright and Jane Standen; Somerset County Museums Service – Stephen Minnitt; Torquay Museum – Ros Palmer; United Kingdom Detecting Finds' Database – Rod Blunt; Western Australian Maritime Museum – Myra Standbury; Wiltshire Heritage Museum – Paul Robinson; and the Worshipful Company of Mercers – Donna Marshall.

Military buttons, subsequently eliminated from this work, were identified by Tim Mole and David Knight, and button collectors Gillian and Alan Meredith helped in other ways – my thanks to them all. Special acknowledgement goes to two of the above mentioned people, Geoff Egan and Paul Robinson, and also to Nick Griffiths, all of whom willingly shared their wisdom and proffered advice, which was gratefully received. Geoff kindly also proof-read the manuscript and wrote the foreword. Paul went out of his way to provide study facilities and a warm welcome at his museum, plus welcoming mugs of coffee. Nick received us into his home, furnished hospitality, and interrupted his busy schedule to expertly draw many of the buttons: he also made me aware of the button- and toggle-like objects depicted on the Camomile Street soldier, toggle-like fasteners worn by a centurion on the Chatsworth relief, Salisbury's feasible connection with Skåne-type jewellery, and much more besides.

Without the help of the aforementioned stalwarts of the Society of Thames Mudlarks, access to a remarkable assemblage of medieval and early post-medieval metal buttons – a representation of which is included herein – would have been denied. My thanks to Ken Bellringer for telling me about these buttons and Vic Housten for allowing examples to be photographed in his home. Geoff Egan brought to my attention a group of 17th-century buttons and a possible cloak-fastener from the River Thames foreshore, London, belonging to a third Mudlark, the aforesaid Tony Pilson. These objects had been drawn by Nick Griffiths: I am indebted to Tony, Geoff and Nick for generously permitting the publication of these illustrations here.

My son Patrick's artistry is responsible for a large proportion of the line drawings, he also photographed a 17th-century button in Torquay Museum, and I am grateful to him for spending many hours at the drawing board. My thanks to master goldsmith Barry Sherlock who advised on technical aspects of his ancient craft: concerning the manufacture of metal buttons, his observations proved invaluable and enabled composite buttons to be described more accurately. I am appreciative of talented

Philippa Foster of 5 D Illustration for not only designing the book and its cover, but also undertaking the typesetting: not once did she complain about this pernickety author who on several occasions added new material because of his inability to say 'THE END.' Finally, I am indebted to my partner Val for accompanying and working alongside me on numerous research trips, and proof-reading the manuscript at various stages of production.

Foreword

In this latest volume Brian Read has turned his attention to one specific category of dress accessories. Metal buttons, medieval and early-modern examples of which are often simply plain and sub-spherical, are not the easiest objects to put into rational series, even with the meagre hints of a chronological framework provided by contemporary illustrations (including sculpture) and the few closely dated archaeological finds. There is more scope with decoration, which very occasionally furnishes unequivocal evidence for the precise date at which an accessory, or even a definable series, was in fashion. Some mistakes are inevitable if one ventures beyond the restricted ambit of finds from firm archaeological sequences. This highly complicated, still developing subject can easily become overwhelmed by reliance on flimsy arguments based on minute detail. In highlighting fixed points while avoiding dogmatic presentation of neat progressions, the present work may be seen as a significant step on the way towards an almost certainly unachievable complete history of early buttons in England. The large number of clear drawings and photographs, brought together in this book with considerable labour, are an essential complement to the text.

While seeking to minimise the speculative dating which has previously bedevilled this subject, this volume presents a considered picture of the current state of knowledge of the range of early buttons up to c.1700, drawing particularly on many previously unpublished detectorists' finds, which over the past thirty years have massively expanded the numbers available for study. Within the broad scope this guide aims to cover, a balance has been struck, exercising caution when deciding whether or not to include less certain items. Even the decision to categorise a particular object is a button has its difficulties – strictly, a button in the modern sense is a *knop* attached at one edge of a textile or leather garment or shoe (or a holder, like a bag) with a corresponding hole in the other edge for closure together - a combination that is only very exceptionally possible to demonstrate for excavated finds. A few prehistoric and Roman items have been included to give a fuller chronological picture where precise function is uncertain.

Buttons have a strange power to fascinate a variety of enthusiasts, from several different viewpoints, out of all proportion to the small scale of the actual objects. This is without doubt the most reliable and accurate guide in its field now available. It will appeal to fashion historians, archaeologists and collectors alike.

Geoff Egan, London 2010

Introduction

Despite an absence of proof that Late Bronze Age folk used buttons on their clothing, button-like objects of bone, cast or sheet bronze, or shale clad with sheet gold or sheet bronze are recorded from archaeologically stratified deposits. From this same period, the leisure-pursuit of metal-detecting has also contributed examples of the metal button-like types. Metal button-like objects attributed to the Iron Age are unrecorded from Britain; however, the British Museum has four from Greece, two of which are nos 7, 8 below: precisely where or how they were used is uncertain. Similarly lacking in confirmed evidence of their specific function are Late Iron Age and Roman period toggle-fasteners and Late Iron Age and Roman period button-and-loop fasteners; again these are known from archaeological contexts and as metal-detector finds.

There is no consensus as to whether metal buttons as we know them today were used in Roman Britain. Metal buttons sometimes classified as Roman, occasionally turn up on Romano-British sites, both as archaeological recoveries and as chance finds by metal-detectorists. However, with one possible exception, no. 25, examination of such buttons indicates they are probably intrusive. Notwithstanding, Romano-British button-like objects are recorded from Britain, see nos 26-27.

Other than within a handful of archaeological reports and books and in metal-detecting books and magazines, little has been written about base-metal toggle-fasteners or button-and-loop fasteners. The same situation applies to precious- or base-metal buttons that antedate AD 1700. Geoff Egan's and Frances Pritchard's invaluable *Dress Accessories c.1150 – c.1450* is the only reference book that deals in some depth with medieval base-metal buttons recovered from reliably dated contexts. Although excellent reference books are available on medieval, Tudor and Stuart dress and jewellery – including elaborate precious-metal and gemstone buttons – they provide scant information about the construction of the buttons.

In Britain, it is perceived that metal-detectorists and archaeologists rarely discover medieval base-metal buttons. It is difficult to understand why not, for in London they have been found on archaeological excavations and the foreshore of the River Thames, especially by members of the Society of Thames Mudlarks. It would be presumptuous to believe that medieval base-metal buttons do not turn up more regularly away from the capital – probably they do, but go unrecognised (as a direct consequence of the first edition of this book, the author has recorded several late medieval metal buttons which are catalogued herein). Accepting that from London at least, confirmed medieval base-metal buttons are not that uncommon, to date, they appear to be limited to a handful of types, some of which are undecorated while others have moulded-in-relief, die-stamped or engraved decoration.

In contrast – undoubtedly reflecting a blossoming in the number of metal buttons used on individual items of clothing (especially doublets and jerkins), combined with more of the populace wearing buttons on their clothing – from sites throughout the country 16th-century base-metal utilitarian buttons are not uncommon, while those attributed to the 17th century are ubiquitous. The enthusiasm of the Mudlarks, who both metal-detect and dig and sieve the Thames foreshore and assist on archaeological sites in London, and metal-detectorists the length and breadth of the land, is largely responsible for a plenitude of base-metal buttons available for study.

Although extremely rare, silver or gold buttons attributed to the late Middle Ages and 16th and 17th centuries are occasionally recovered, with shipwrecks, particularly the Spanish vessel the *Girona*, having provided beautiful examples. Perhaps the most exquisite are gold, silver or silver-gilt filigree buttons, with *c.* late 16th-century silver or silver-gilt examples recorded as fortuitous finds by metal-detectorists. Pseudo filigree buttons of the same period, which are just as scarce, are perhaps less pleasing to the eye. The repealing of sumptuary law in England in 1604 meant the lifting of the prohibition on wearing precious-metal jewellery, including buttons, by the lower classes.

Some 16th-century base-metal buttons are rather plain, although the fronts of many have engraved or moulded-in-relief decoration. Several types of 16th-century base-metal button characteristically have their fronts black painted (possibly bituminous) or nielloed. Decoration of base-metal buttons increased during the 17th century and moulded-in-relief, engraved, punched or die-stamped, openwork, gilded, silvered, tinned, black painted, champlevé enamel and appliqué techniques were employed. A ubiquitous class of cast base-metal button of this period, invariably well made and highly decorative, perhaps formed one half of a cloak-clasp, although its use as a true button is also likely. Another frequently-found late 17th-century (and other periods) 'half of', and often mistaken for a button, is the cufflink. These constitute a separate area of study which falls outside of the parameters of this book, nonetheless, for comparison, several examples are catalogued here.

Accurately dating a metal button found in an unstratified context is invariably fraught with difficulty, for like many classes of everyday domestic small metalwork, some styles of metal button appear to have remained unchanged for centuries. Contemporary inventories and wills indicate that silver buttons were often intrinsically valuable heirlooms, being bequeathed from generation to generation, thereby further exacerbating the problem of assigning a date of manufacture. It is reasonable to assume that even mundane utilitarian base-metal buttons sometimes had an extended period of use by being removed from old clothing and reused on new.

Buttons – often very elaborate and some of which are obviously purely decorative – depicted on contemporary paintings, effigies and monumental brasses rarely show more than one perspective: notable exceptions are the Fountain statue (the Hansel), Nuremberg, *c.*1380, in the Germanisches Nationalmuseum, Nuremburg; a German 1472 anonymous oil painting of a Town Secretary in the Statens Museum fr Kunst, Copenhagen; and a *c.*1560-65 portrait in oil of Robert Dudley, Earl of Leicester, K.G., by Steven van der Meulen (the Wallace Collection, London). Therefore, one can only speculate on precisely how the buttons were made, which means they usually prove of limited use in dating similar examples found out of context (be circumspect, for some early post-medieval buttons, which in contemporary depictions appear to be metal, are actually wooden cores clad with textile or gold or silver braid, and are known as passementerie buttons). These depictions – particularly of the 16th and 17th centuries – invariably are of royalty, the aristocracy or other notables; the opulence of their dress, especially gemstones and precious-metal jewellery and buttons, is making a statement of just how wealthy they are. Commoners were rarely immortalised this way, which means a deficit of knowledge about their dress accessories, particularly quantities of, and precisely where different types of utilitarian base-metal buttons were worn. In the absence of a reliable attribution from a stratified context, extant dress or contemporary depiction, one is restricted to dating a button stylistically; however, longevity should be borne in mind.

Caution is advised with blindly accepting the dating of metal buttons recorded in some archaeological reports: the more often than not poor standard of illustration and lack of textual precise constructional description sometimes precludes the reader from making a reasoned judgement. When an illustration is decipherable, a button's style sometimes suggests it is perhaps intrusive. These caveats also apply to early metal-buttons published in metal-detecting periodicals and books. Notwithstanding, although dating is often suspect, the very act of publishing is commendable, for it encourages discussion of these humble yet fascinating dress-fasteners of yesteryear among a wider audience.

For buttons of any metal that have one, the type of shank may be an aid to dating; however, this may lead one totally in the wrong direction, therefore be circumspect. Another confusing, but crucial, aspect of a metal button's character is the number of separate parts it comprises: herein the author has elected to deviate slightly from the norm and, other than appliqués, filigree and granules, each individual piece (including the shank) is counted as one. Cast solid-metal buttons with integral shanks are described as one-piece, whilst cast, die-stamped sheet, repoussé, filigree, or non-filigree drawn wire metal buttons that have between two and five separate components are classed as composite. These are differentiated both textually and illustratively.

After centuries of burial in the ground, many leaden objects are recovered in a stable condition, having an even off-white patina, whilst others are disintegrating due to corrosion. The degree of corrosion is dependant on several factors, all of which are time related: whether the soil is acid or alkaline, the quantity and type of agrochemical or other chemical in the soil, the degree of exposure to the atmosphere (i.e. oxygen), mechanical damage (e.g. farm machinery), and, if it's an alloy, its precise composition. Acid soil, agro and other chemicals, oxygen and mechanical damage all play a part in accelerating decomposition. These same destructive causes apply equally to objects made from copper and its alloys. The crumbly nature of some excavated leaden buttons suggests they are made from eutectic pewter, i.e. the alloy contains a higher proportion of tin (typically about three parts of tin to two parts of lead), whilst others are found in a sound condition, which indicates they are lead or have a high-lead content. However, determining the precise composition of a base-metal alloy is impossible without scientific analysis, therefore – apart from buttons with a known composition from the collection of the Museum of London – herein, copper and its alloys are described as copper alloy, and lead and its alloys as lead/tin alloy (several 17th-century buttons described here are perhaps tombac, an alloy of copper and zinc). Applied surface-decoration is usually easier to identify, and here standard terminology is used, i.e. gilt, gilded, gilding, gold-leaf, white-metal coating (either tin or silver), enamel, enamelled and painted. Caution is advised though, for not all that glitters is gold: base-metal artefacts, especially lead/tin alloy, which have lain for decades in undisturbed waterlogged oxygen-free conditions, frequently look gilded which is actually an adventitious sulphide deposit known as anaerobic gilding (commonly called nature's gilding). A combination of oxygen exclusion, and abrasion caused by movement, often replaces patina with a brassy or coppery lustre on copper-alloy buttons retrieved from watercourses or the seashore. Note buttons from these two categories found in the Thames foreshore mud.

The paucity of documentation available to button collectors and dealers, museum curatorial staff, dress historians, archaeologists or metal-detectorists, concerning any of these classes of metal fastener, influenced the writing and publication of this book, which makes no claim for being definitive (different types of button, and decoration on, are continually coming to light). With several exceptions, the early metal button-like objects, toggle-fasteners, button-and-loop fasteners and buttons described were all found in England. Metal-detectorists and the Thames Mudlarks are responsible for the greater proportion of this assemblage, however, examples – some of which are also metal-detecting discoveries – from the collections' of several museums are included. Find-spots are denoted in italics: to protect the confidentiality of landowners and sites, the omission of precise provenance information is deliberate.

Every effort has been made to ensure accuracy with the dating, construction and identification of objects catalogued here, based upon the information to hand at the time of writing. The inclusion of new material, coupled with several deletions of old, and the correction of errors, inevitably affected the layout of this second edition, it is hoped, for the better. As before, observations made by persons mentioned in the Acknowledgement were considered, but any misidentifications, inaccuracies or errors that do remain are solely the responsibility of the author.

1: Late Bronze Age Buttons

These button-like objects have either an undrilled simple loop, a drilled shank or stitching-holes on their backs, suggesting they were affixed to a secondary object, but whether for fastening clothing is unknown. All are possibly solely decorative.

Cast one-piece copper-alloy button-like objects with integral undrilled shanks
Nos 1 and 3 are from secure *c.*900 – *c.*600 BC archaeological contexts. Decoration is moulded-in-relief.

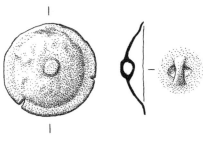

1. Discoidal; convex front; concave back; a central flat-topped boss and a slightly downwards canted rim, split in two places; the oval-section simple looped shank has signs of wear in the centre; D 30mm; shank L 2mm. *Switzerland.*

2. Discoidal; convex front; concave back; a central flat-topped boss within two concentric circles, the outer one forming a downwards canted rim; a circular-section simple looped shank; D 22mm; shank L 5.5mm. *South-East Dorset.*

3. Discoidal; solid; convex front; flat back; a central flat-topped boss within two concentric circles; a downwards canted rim; incomplete, two offset circular-section projections on the back imply a broken-off shank; D 21mm. *Wiltshire.* WHM, acc. no. 1994.294.5.

Cast one-piece copper-alloy button-like object with integral drilled shank
Found in association with a scatter of confirmed Late Bronze Age metalwork, however, apart from the curious projections, its similarity with nos 28-32, 38, 40 suggests it is perhaps intrusive and of medieval or post-medieval date. Nonetheless, here it is tentatively attributed to the Late Bronze Age. Probable moulded recessed decoration.

4. Discoidal; flat; a central circle-and-dot motif within a circle; a wedge-section trapezoid shank with signs of wear on one side. A curiosity, for the edge of the disc is smooth, but in one place it is slightly ragged and torn and from two places there is a broken protrusion; the damaged section possibly represents another missing protrusion. Whether these protrusions are remnants of some form of peripheral ornamental openwork or casting sprues is uncertain; D 24mm; shank L 12mm. *South-East Dorset.*

Probable composite two-piece die-stamped or repoussé sheet copper-alloy and shale button-like object with indeterminate shank
From a secure *c*.900b– *c*.600 BC archaeological context. Decoration is die-stamped or repoussé.

5. Discoidal; convex; a central boss within a circle; the rim cants upwards and then laterally; incomplete, section of front broken off, separate shale back missing; D 23mm. *Wiltshire.* WHM, acc. no. unknown.

Composite three-piece sheet gold and shale button-like object with (?) stitching-holes
From a secure *c*.900 – *c*.600 BC archaeological context. Probable impressed decoration.

6. Discoidal; conical front; flat back; a cone of solid shale with a recessed rim; clad overall with two pieces of sheet gold; two stitching-holes in the back. The shale presumably has a cavity to allow passage of a cord or thong. Both the front and the back are decorated with bands of multiple concentric circles and dots; the front has a ragged-edged overlapped joint, whilst front and back are seamed together; D 41mm. *Wiltshire*. After Annabelle and Simpson 1964, no. 181.

2: Early Iron Age Buttons

Composite two-piece die-stamped or hammered sheet copper-alloy button-like objects with probable separate undrilled shanks

Whether true buttons used for fastening clothing, or how or where they would have been worn, is unknown. Both are made from sheet metal, either hammered or die-stamped into a dished shell. Corrosion precludes determining how the shanks are affixed; both are probably separate and inserted through holes in the shell and soldered on the inside. Both are *c*.500 BC, and decoration is impressed-in-relief.

7.   Discoidal; concave front; convex back; a slightly convex terracotta insert depicting a facing Gorgan's mask; a circular-section simple looped shank; D 19mm; shank L 6.5mm. *Greece*. Precise provenance and circumstances of discovery unknown. BM, acc. no. GR 1959. 7-20 3.

8.   Discoidal; concave front; convex back; a flat terracotta insert depicting a facing Gorgan's mask; a circular-section simple looped shank; D 20mm; shank L 3mm. Similar to no. 7, but deeper dished and a smaller shank. *Greece*. Precise provenance and circumstances of discovery unknown. BM, acc. no. GR 1959. 7-20 4.

3: Late Iron Age and Roman Period Toggle-Fasteners

Cast one-piece copper-alloy toggle-fasteners with integral undrilled shanks or shank-less

Whether these invariably substantial objects were solely for fastening clothing is uncertain. Use other than with dress is a possibility. Two basic forms are noted, one of which has a loop and the other loop-less. A horizontal toggle-like fastener is shown on the *paenula* (military cloak) of the Camomile Street soldier, fig 1, as are four on the Chatsworth relief centurion. The former appears to be inserted through a pair of loops, one attached to each hem, which, if correct, seems a precarious securing. This toggle is perhaps a dumb-bell type, like no. 14, rather than one with a loop. Nos 9-12 are *c.*50 BC – *c.*AD 100. Nos 13 and 14 are Roman period, probably post-43 AD. Unless otherwise stated, decoration is moulded-in-relief.

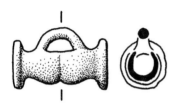

9. Cylindrical; solid; the body has a central narrow waist, each side of which expands and then narrows into a waist before expanding into a transverse discoidal end; a circular-section simple looped shank; L 25mm; D 13mm; shank L *c.*5mm. *South-West Wiltshire.*

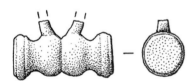

10. Cylindrical; solid; the body has a central narrow waist, each side of which expands and then narrows into a waist before expanding into a transverse discoidal end; a peripheral ridge around each end; incomplete, a probable circular-section simple looped shank broken off; L 26mm; D 12mm; shank remnant L 4mm. *North Dorset.*

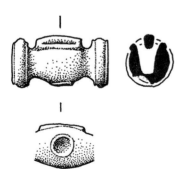

11. Cylindrical; solid; the body has an expanded centre, each side of which narrows into a waist before expanding into a transverse discoidal end; a peripheral ridge around each end; a wedge-section shank, the rectangular loop of which projects slightly above the body; below the shank is a lipped recess which allows access for a strap, thong or cord; opposite the shank, and slightly off centre, is a collet which perhaps held an appliqué; L 25mm; D 12mm; shank L 3mm. *North Dorset.*

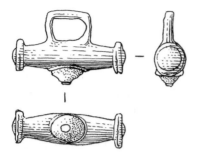

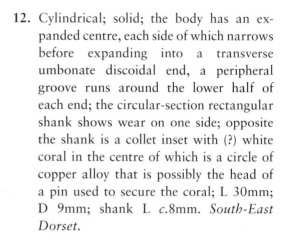

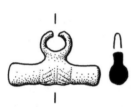

12. Cylindrical; solid; the body has an expanded centre, each side of which narrows before expanding into a transverse umbonate discoidal end, a peripheral groove runs around the lower half of each end; the circular-section rectangular shank shows wear on one side; opposite the shank is a collet inset with (?) white coral in the centre of which is a circle of copper alloy that is possibly the head of a pin used to secure the coral; L 30mm; D 9mm; shank L *c.*8mm. *South-East Dorset.*

13. Cylindrical; solid; the body has an expanded centre, each side of which narrows slightly before expanding into a transverse slightly convex discoidal end; two parallel bands of oblique lines run from each side of the incomplete circular-section simple looped shank; D 6mm; L 25mm; shank L *c.*8mm. *South-East Dorset.*

14. Cylindrical; solid; a narrow body that expands at each end into a globule, thereby forming a dumb-bell shape; L 23mm; D of globules 11mm. An unclassified curiosity: others of similar form in copper alloy are known and an example made of bone is recorded in FFRA, no. 365. *Cambridgeshire.*

4: Late Iron Age and Roman Period Button-and-Loop Fasteners

Cast one-piece copper-alloy button-and-loop fasteners with integral undrilled shanks
Ten classes and several unclassified of these usually robust objects are recorded (see Wild 1970). Their precise function is uncertain; whether for fastening clothing, perhaps capes, is a possibility, while use with horse-harness or sword-scabbards are other theories. Notwithstanding, the Camomile Street soldier's cape has two – originally possibly three – discoidal fasteners that may well be of the button-and-loop type, fig. 1, (further comment is made later). The distinction between button-and-loop fasteners and the preceding toggle-fasteners is somewhat abstruse, for toggle-fasteners may have a loop, and button-and-loop fasteners are known with a toggle. Although a Late Iron Age attribution is possible for some, *c*.AD 70–150 is generally accepted. Concerning no. 24, it is uncertain whether this is a button-and-loop fastener, toggle-fastener or a pestle from a Roman period (?) cosmetic pestle-and-mortar (pers. comm. Dr Ralph Jackson, forthcoming). Despite an apparent lack of surviving costume with such an item still attached, or depiction in a contemporary portrait, a suggestion that it is a possible fastener for a Victorian coachman's cloak is doubtful. Interestingly, members of the Thames Mudlarks have not recovered such an object from the Thames foreshore or spoilheap in London and as most of their finds are post 14th century (pers. comm. Tony Pilson), it suggests they may antecede this date. Unless otherwise stated, decoration is moulded-in-relief.

15. Twin ring-headed; flat front; flat back; undecorated; the rings are D-section; a circular-section trapezoid shank projects at 45° immediately from the V of the circles and then runs on the same plane as the head; rings D 14mm; shank L 17mm. Wild's Class I variant. *Chilterns*.

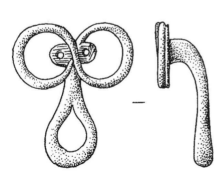

16. Twin-ring-headed; flat front; the rings form a figure-of-eight and each ring is circular-section with an integral circumferential angular-edged step at the back; a pierced lobe, both exhibiting (?) tool marks, each side where the rings cross; a circular-section trapezoid shank projects at 90° immediately from the V of the circles and then runs on the same plane as the head; rings D 17.5mm; shank L 34mm. Wild's Class 1 variant. *South Gloucestershire*.

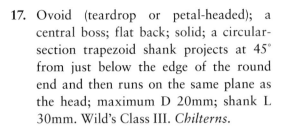

17. Ovoid (teardrop or petal-headed); a central boss; flat back; solid; a circular-section trapezoid shank projects at 45° from just below the edge of the round end and then runs on the same plane as the head; maximum D 20mm; shank L 30mm. Wild's Class III. *Chilterns.*

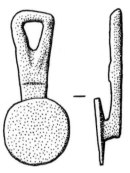

18. Discoidal; flat front; slightly concave back; undecorated; a rectangular-section trapezoid shank with a transverse ridge projects at 90° from the centre of the back and then runs on the same plane as the head; D 18mm; shank L *c.*35mm. Wild's Class Vc. *South-East Dorset.*

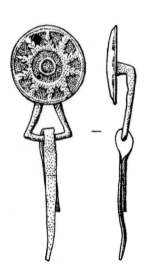

19. Discoidal; convex front; flat back; a central annulet within an elaborate ten-rayed sunburst; a lateral rim; sporadic dark blue champlevé enamel field; a rectangular-section trapezoid shank projects at 90° from the centre of the back and then runs on the same plane as the head; a separate sheet bronze 'split pin' hangs from the shank, one point broken off; D 21mm; shank L *c.*21mm; split pin L 35mm. This is the only button-and-loop fastener in the known record with such a 'split pin': whether this flimsy survival represents a method of attachment or a secondary use is uncertain. Wild's Class Vb. *South-East Lincolnshire.*

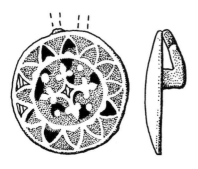

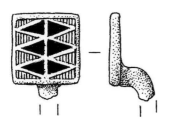

20. Discoidal; convex front; flat back; a cross flory, four pellets and a circle within an elaborate 12-rayed sunburst; sporadic red and dark blue champlevé enamel; incomplete, a rectangular-section probable trapezoid shank, with the end of the loop broken off, projects at 90° from the centre of the back and would have run on the same plane as the head; D 32mm; shank remnant L *c.*15mm. Wild's Class Vb. *South Somerset.*

21. Square; flat; a panel of red and dark-blue champlevé enamelled triangles; incomplete, a circular-section shank, with the end of the loop broken off, projects at 90° from the back and then turns on the same plane as the head; L 19.5mm; W 18.5mm; shank remnant L 9mm; Wild's Class VIa. *East Devon.*

22. Cylindrical; bulbous terminals with flattened recessed ends, one with a vestige of red champlevé enamel, the centre is expanded and has a circumferential rib each side of the loop juncture; a circular-section trapezoid shank projects at 90° and then runs on the same plane as the head; D 10mm; shank L 24mm. Wild's Class IX describes this type as 'fastener with bar for a shank', however, his catalogue cites two examples as 'bar head' with 'square' and 'round loop' respectively, a discrepancy pointed out by Hattatt 1989. *Chilterns.*

23. Shallow plano-convex; undecorated; a rectangular-section shank projects at 90° from the centre of the back and then runs on the same plane before turning again at 90°, forming a cranked shape; before joining a circular-section transverse toggle; dimension unknown. A curiosity that is perhaps a variant of Wild's Class IX. *Chilterns.*

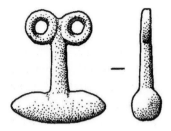

24. Twin-ring-headed; all parts circular-section and in the same plane; a transverse toggle with pointed ends; L 26mm; W 23mm. Perhaps a variant of Wild's Class IX (see comment above). *East Devon.*

5: Roman Period Buttons

Archaeology has not produced any conclusive evidence that true buttons were used as dress-fasteners in Roman Britain or indeed elsewhere in the Empire. If such buttons were current at that time, it is reasonable to believe that they would be visible on some of the wealth of extant contemporary depictions of dress on statues, frescoes and mosaics. Notwithstanding, as mentioned earlier, the Camomile Street soldier's cape has a pair of discoidal fasteners, which are perhaps button-and-loop, but equally they could be true buttons. Roman period button-like objects are recorded, but whether they were used as dress-fasteners is unknown. Here, three examples are described, and their possible function is clearly evident.

Cast one-piece copper-alloy button-like object with integral drilled shank
Stylistically, especially the type of shank, it is similar to cast one-piece buttons attributable to the 15th – 17th century and even the possible Late Bronze Age button-like object no. 4. Uncertainty surrounds whether the find-spot was a stratified deposit, therefore, here, it is tentatively attributed as 1st – 4th century AD. Decoration is either recessed-moulded or incised.

25. Discoidal; shallow convex front; flat back; a border of oblique V-shaped grooves; a rectangular-section rectangular shank with an offset hole; maximum D 23mm; shank L 17.5mm. *Great Chesters Roman Fort, Hadrian's Wall, Northumberland,* acc. no. CH 851. Found around the beginning of the 20th century. Budge 1903 describes this object as a 'stud'.

Cast one-piece copper-alloy button-like object with possible integral drilled shank
Probably 1st century AD, perhaps post-43 AD.

26. Shallow biconvex; two in-line central pellets, each with vestiges of dark blue champlevé enamel, within two confronted crescents and two confronted sub-triangles, each with a vestige of red champlevé enamel; incomplete, a rectangular-section possible trapezoid shank with the end broken off; D 28.79mm; shank remnant L 6.5mm. *Mid-Kent.* PAS KENT-D99B02.

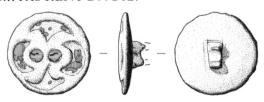

11

Cast one-piece copper-alloy button-like object with integral undrilled shank
Circa 1st – *c*.4th century AD. Decoration is engraved or punched.

27. Discoidal; flat; central two concentric circles within four evenly spaced circle-and-dot motifs; incomplete, a small section of the serrated slightly chamfered edge is broken off; a circular-section simple looped shank; D 37mm; shank L 8mm. *South-East Dorset*. After Read 2001, no. 22, described as a 'strap slide', which although possible is unlikely unless the strap was extremely narrow. The shank seems more suitable for stitching to or inserting through thick woollen textile or leather. Note the similarity of the shank with no. 1.

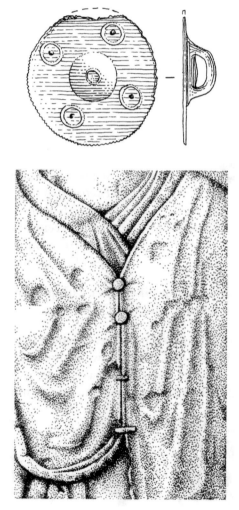

Fig 1. The Camomile Street soldier, detail of button-like
and toggle-like fasteners, in the MoL.

12

6: Medieval Buttons

Although contemporary medieval depictions of dress are of limited use for understanding how their buttons were made, often one can see where such fasteners were attached; for example, fronts of tunics, capes and dresses; and on sleeves, cuffs and collars. However, precisely where some types of medieval button were worn on dress is uncertain.

Cast one-piece copper-alloy buttons with integral drilled shanks
All are possibly *c*.15th century, although the several styles of shank on this type of button perhaps continued beyond the late 17th century. Nos 29-34, 38-39, 46, 51 are from a *c*.15th-century context. Note the similarity of shanks with nos 4 and 25. Unless otherwise stated, decoration is recessed, either punched, engraved, compass inscribed and probably moulded.

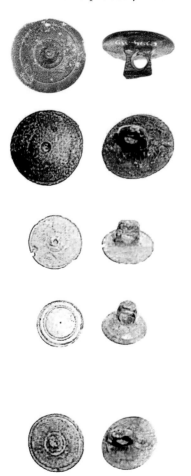

28. Shallow plano-convex; solid; a central pit within three concentric circles; a rectangular-section sub-rectangular shank; D 21mm; shank L *c*.7mm. *Wiltshire*. WHM, acc. no. 1994.119.

29. Shallow plano-convex; solid; a central pit within two concentric circles; a rectangular-section rounded shank; (note anaerobic gilding). *River Thames foreshore, London.*

30. Shallow plano-convex; solid; a central pit within three concentric circles; a rectangular-section trapezoid shank; (note anaerobic gilding). *River Thames foreshore, London.*

31. Shallow plano-convex; solid; a central pit within three concentric circles; an upwards canted border; a rectangular-section rectangular shank with chamfered edges; (note anaerobic gilding). *River Thames foreshore, London.*

32. Shallow plano-convex; solid; a central pit within two concentric circles; a rectangular-section sub-rectangular shank; (note anaerobic gilding). *River Thames foreshore, London.*

13

33. Shallow plano-convex; solid; a central pellet and pits forming a quatrefoil with pointed sepals within a dotted circle; a rectangular-section rounded shank; (note anaerobic gilding). *River Thames foreshore, London.*

34. Shallow plano-convex; solid; dots forming a saltire with two concentric circles in the centre and one in each quadrant, within a circle of dots; a rectangular-section rounded shank; (note anaerobic gilding). *River Thames foreshore, London.*

35. Shallow plano-convex; solid; a septfoil formed from circles; all of which are recessed and filled or partially filled with a dull, now silvery substance that is perhaps niello, within a circle; a rectangular-section rounded shank; D 12mm; shank L 5.5mm. *East Devon.*

36. Shallow plano-convex; solid; circles forming a sexfoil with sepals, within a circle; a rectangular-section rounded shank; D 13mm; shank L 5mm. This decoration remained popular on buttons well into the 17th century. *South Devon.* After Read 1995, no. 987.

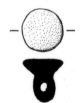

37. Plano-convex, solid; undecorated; a rectangular-section rounded shank; D 17mm; shank L 10.5mm. *West Yorkshire.* UKDFD 9285.

38. Discoidal; concave front; slightly convex back; four concentric circles; a rectangular-section rounded shank. *River Thames foreshore, London.*

14

Fig 2. Monumental brass of Robert Wyvil, Bishop of Salisbury (died 1375), in Salisbury Cathedral. The Bishop's champion is depicted standing outside the door of Sherborne Castle, Dorset. Note plain discoidal buttons on the champion's tunic. After Kite 1860.

15

39. Discoidal; shallow concave front; shallow convex back; two concentric circles, hatching in the centre and the border; a lateral rim; a rectangular-section trapezoid shank. *River Thames foreshore, London.*

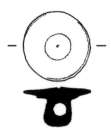

40. Shallow biconvex; solid; a central pit within two concentric circles; a rectangular-section trapezoid shank; D 18mm; shank L 7mm. *South Somerset.*

41. Shallow biconvex; solid; a sexfoil infilled with circle-and-dot motifs within two concentric circles; a rectangular-section rounded shank; D 13mm; shank L 5mm. *East Devon.*

42. Shallow biconvex; solid; a sexfoil with circle-and-dot motifs within two concentric circles; a rectangular-section rounded shank; D 13.5mm; shank L 5.5mm. *North Dorset.*

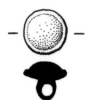

43. Shallow biconvex; solid; undecorated; upwards canted edge with a lateral rim; a rectangular-section rounded shank; D 11mm; shank L 5.5mm. *South-West Wiltshire.*

44. Discoidal; shallow concave front; shallow convex back; a central pit and four arcs, behind each are curvilinear bands of zigzags; radiating lines in the two quadrants; a rectangular-section trapezoid shank; D 18.2mm; shank L 7.5mm. *South-West Wiltshire.*

45. Discoidal; shallow concave front; shallow convex back; a central pit and two chevrons formed from voided bands of addorsed triangles, two angles thus formed have a voided band of addorsed triangles and the other two angles, radiating and oblique lines, all within a circle; a rectangular-section trapezoid shank; D 19.5mm; shank L 9mm. *North Dorset.*

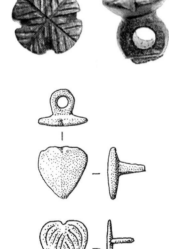

46. Quatrefoil; solid; a plain cross with oblique lines in each quadrant, resembles four veined-leaves; a rectangular-section rounded shank. *River Thames foreshore, London.*

47. Heart-shaped; shallow biconvex; solid; undecorated; a rectangular-section trapezoid shank; L 19mm; W 18mm; shank L 10mm. *South Somerset.*

48. Heart-shaped; shallow biconvex; solid; shape and decoration resembles a pital leaf; a circular-section rounded shank; L 16mm; W 14.5mm; shank L 6.5mm. *South-West Dorset.*

49. Heart-shaped; concave front; convex back; linear decoration forming a heart within a hatched border; a rectangular-section rounded shank; L 18; W 17mm; shank L 6.5mm. *Kent.* UKDFD 16037.

50. Heart-shaped; concave front; convex back; a heart with a median line within a hatched border; a rectangular-section rounded shank. *East Hertfordshire.*

17

51. Sub-rectangular; solid; convex front; flat back; a moulded-in-relief quatrefoil or a fleur-de-lis – the latter representing the purity of the Virgin Mary; a circular-section oval shank; (note anaerobic gilding). *River Thames foreshore, London.*

Fig 3. Monumental effigy of perhaps Sir Richard de Willoughby (Edward III period), in Willoughby Church, Nottinghamshire. Note plain discoidal buttons on the costume. After Stothard 1817.

Cast one-piece copper-alloy buttons with integral undrilled shanks
The attribution of no. 52 is uncertain, but possibly found in association with pottery dating to *c*.1200–30. Nos 53, 54-55 are perhaps *c*.14th – *c*.16th century. Unless otherwise stated, decoration is moulded-in-relief.

52. Biconvex; solid; high-tin bronze; raised rim; undecorated; a mould line on the back; incomplete, probable circular-section simple looped shank broken off, slight damage on the front; D 13mm; shank remnant L *c*.5mm. *Billingsgate, London.* MoL, BIG82 2482 (2560). After Egan and Pritchard 1991, no. 1383.

53. Shallow biconvex; solid; a sexfoil with circle-and-dot motif in each petal and in the centre within a circle; incomplete, a circular-section short stem, circular-section simple loop broken off; D 12.5mm; shank remnant L 3mm. *South Somerset.*

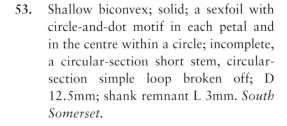

54. Discoidal; solid; convex front; flat back; a sexfoil with a circle-and-dot motif in each leaf and in the centre within two concentric circles formed from zigzags; a lateral rim; a circular-section simple looped shank; D 20mm; shank L 7.5mm. *North Dorset.*

55. Discoidal; flat; a quatrefoil formed from five circle-and-dot motifs, the outer ones within a crescent; pointed sepals; within a circle of tiny annulets; incomplete, remnant of a circular-section simple looped shank; D 15.5mm; shank remnant L 3.5mm. *North Dorset.*

19

Fig 4. A monumental brass of Sir Miles (died 1344) and Lady Joan de Stapleton, formerly in Ingham Church, Norfolk. Note plain discoidal buttons on the front, and spherical buttons on the sleeves and cuffs of the lady's dress. After Stothard 1817.

Cast one-piece lead/tin alloy or tin buttons with integral undrilled shanks
Some of these buttons have archaeological secure dates: no. 69 *c.*1200-30; nos 70-72 *c.*1230-60; no. 73 *c.*1350 – *c.*1400; no. 74 *c.*1400-50 and no. 75 *c.*1270 – *c.*1350. Nos 56, 58-62, 64-67 are from a *c.*15th-century context, however, nos 59-61 could be as early as the 13th century, whilst nos 58, 62-67 are possibly 14th century. Nos 57, 63 are possibly 14th century. The attribution of no. 68, which is possibly not a button, is uncertain, but perhaps *c.*15th century. Nos 76-77 are *c.* late 13th – *c.*15th century. Unless otherwise stated, decoration is moulded-in-relief.

56. Discoidal; flat; a facing human mask, possibly representing Christ or St John the Baptist, although St John is usually shown nimbate; a circular-section simple looped shank; a mould line on the back; (note anaerobic gilding). *River Thames foreshore, London.*

57. Discoidal; flat; a crowned facing mask of a king with (?) long hair, moustache and beard, within a beaded circle; a mould line on the back; a circular-section simple looped shank; slightly abraded; D 15.8mm; shank L 5.3mm. *Gloucestershire.*

58. Discoidal; flat; either the cross of the Knights Hospitallers or St George within a shield and a beaded circle; a mould line on the back; incomplete, a probable circular-section simple looped shank broken off; (note anaerobic gilding). *River Thames foreshore, London.*

59. Discoidal; flat; a fleur-de-lis – representing the purity of the Virgin Mary – within a circle; a circular-section simple looped shank; a mould line on the back and two casting sprues on the edge; (note anaerobic gilding). *River Thames foreshore, London.*

60. Discoidal; flat; a fleur-de-lis – representing the purity of the Virgin Mary – within a beaded circle; a mould line on the back; a circular-section simple looped shank; projection on side is damage; D 15mm; shank L 3.2mm. *River Thames foreshore, London.*

61. Discoidal; flat; a quatrefoil with a plain cross in each petal and a central annulet, a pellet between each petal; within a beaded circle; a circular-section simple looped shank; a mould line on the back. *River Thames foreshore, London.*

62. Discoidal; flat; a W within a circle; a circular-section simple looped shank; a mould line on the back; (note anaerobic gilding). *River Thames foreshore, London.*

63. Discoidal; flat; a black-letter A within a circle; a circular-section simple looped shank; a mould line on the back; D 16.8mm; shank L 10mm. *North Yorkshire.*

64. Discoidal; flat; a sexfoil formed from a central pellet and alternate pellets and annulets within a circle; a circular-section simple looped shank; a mould line on the back and a casting sprue on the edge; (note anaerobic gilding). *River Thames foreshore, London.*

65. Discoidal; flat; a saltire with a central pellet, a trefoil of pellets in each quadrant, all within a circle; two circular holes are probably blow-holes caused by air bubbles trapped in the mould; a circular-section stem and a circular-section loop; a mould line on the back and two casting sprues on the edge; (note anaerobic gilding). *River Thames foreshore, London.*

66. Discoidal; flat; a saltire with a pellet in each quadrant; a circular-section stem and a circular-section loop; a mould line on the back; (note anaerobic gilding). *River Thames foreshore, London.*

67. Plano-convex; probably hollow; a five-petalled rose within a beaded border; a circular-section simple looped shank; a mould line on the back; back dented; (note anaerobic gilding). *River Thames foreshore, London.*

68. Sub-discoidal; flat; a saltire with pellets on the arms within a pelleted and cross-hatched border; slightly downwards canted rim; sporadic gold-leaf in three quadrants; a mould line and a small circular pit on the back; incomplete, shank broken off; D 28mm. *Wiltshire*. WHM, acc. no. 92.1980.

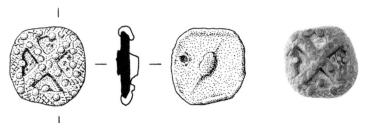

69. Biconvex; solid; tin; a beaded rim; a mould line on the back; possible circular-section simple looped shank; D 8.5mm; shank L *c*.6.5mm. *Billingsgate, London*. MoL, BIG82 3403 (3204). After Egan and Pritchard 1991, no. 1376.

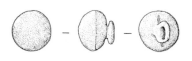

70. Biconvex; solid; tin; undecorated; a mould line on the back; possible circular-section simple looped shank; D 11mm; shank L *c*.4.5mm. *Billingsgate, London*. MoL, BIG82 2649 (2591). After Egan and Pritchard 1991, no. 1377.

71. Biconvex; solid; pewter; abraded front; a border of zigzags or dots; a mould line on the back; a circular-section simple looped shank; D 12.5mm; shank L *c*.7cm. *Billingsgate, London*. MoL, BIG82 2508 (2853). After Egan and Pritchard 1991, no. 1378.

72. Biconvex; solid; tin; a central plain collet inset with a yellow glass stone within a beaded border; a mould line on the back; a possible circular-section long stem and a circular-section loop; D 12mm; shank L 7.5mm. *Billingsgate, London*. MoL, BIG82 2338 (2745). After Egan and Pritchard 1991, no. 1379.

73. Biconvex; solid; high-tin pewter; a central circular indentation may have originally held a separate stone set within a beaded border; a mould line on the back; abraded; incomplete, a possible circular-section shank broken off; D 13mm. *Billingsgate Lorry Park, London.* MoL, BWB83 4924 (373). After Egan and Pritchard 1991, no. 1380.

74. Shallow plano-convex; solid; tin; a six-pointed wavy star (possibly representing a royalist or Yorkist badge) and a central pellet within a raised rim bordered with pellets; a mould line on the back and the shank; a circular-section short stem and a circular-section simple loop; D 12mm; shank L *c*.9mm. *Swan Lane, Upper Thames Street, London.* MoL, SWA81 1782 (2103). After Egan and Pritchard 1991, no. 1381.

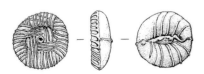

75. Possibly originally spherical; tin; hollow; the front has a central annulet within four fields of linear decoration, each field linked transversely at 90°; the back has five transverse bands; incomplete, shank broken off, squashed and fractured on the back; the fracture is on the mould line and perhaps represents two blow-holes; D 18mm. *Billingsgate Lorry Park, London.* MoL, BWB83 2121 (290). After Egan and Pritchard 1991, no. 1396.

76. Biconvex; solid; undecorated; a mould line on the back and around the circumference; incomplete, a circular-section shank broken off; D 10mm. *South-West Wiltshire.*

77. Discoidal; solid; conical front, flat back; undecorated; a circular-section simple looped shank; D 11mm; shank L 3.5mm. *Cambridgeshire.*

Fig 5. Monumental brass of Sir John de Northwood (died 1337), and his wife Joan de Badlesmere, in Minster Church, Isle of Sheppey, Kent. Note plain possible spherical buttons on the front of the lady's partially unbuttoned travelling hood. After Stothard 1817.

Composite two-piece cast copper-alloy buttons with separate embedded drawn copper-alloy wire shanks

All are perhaps *c.* late13ᵗʰ – *c.* mid-14ᵗʰ century, although this style of plain button continued into at least the early post-medieval period. Identical buttons are recorded from Billingsgate, London *c.*14ᵗʰ-century archaeological deposits. More often than not these buttons are high-tin content or have a polished tin coating, both of which may create an appearance of silver. Examples of silver possibly exist. All are undecorated. Also recorded with drawn iron wire shanks.

78. Spherical; solid; polished tinned surface; a circular-section simple looped shank; D 8mm; shank L 5mm. *South-West Wiltshire.*

79. Biconvex; solid; polished tinned surface; a circular-section simple looped shank; D 10mm; shank L 5mm. *South-East Lincolnshire.*

80. Biconvex; solid; polished tinned surface; a circular-section simple looped shank; D 8mm; shank L 4mm. *South Somerset.*

81. Biconvex; solid; polished tin surface; a circular-section simple looped shank; D 9mm; shank L 2.5mm. *South-West Wiltshire.*

82. Biconvex; solid; polished tin surface; a circular-section simple looped shank; D 7.5mm; shank L 3mm. *South-West Wiltshire.*

83. Biconvex; solid; polished tin surface; a circular-section simple looped shank; D 9mm; shank L 2.5 *South-West Wiltshire.*

84. Biconvex; solid; polished tin surface; a circular-section simple looped shank; D 10mm; shank L 4mm. *South-West Wiltshire.*

85. Biconvex; solid; polished tin surface; a circular-section simple looped shank; D 11mm; shank L 2.5mm. *South-West Wiltshire.*

86. Biconvex; solid; polished tin surface; a circular-section simple looped shank; D 12mm; shank L 3.8mm. *South-West Wiltshire.*

87. Biconvex; solid; polished tin surface; a circular-section simple looped shank; D 12mm; shank L 5mm. *Cambridgeshire.*

88. Discoidal; solid; conical front; flat back; a circular-section simple looped shank; D 8mm; shank L 5mm. *South Somerset.*

Composite two-piece cast lead/tin alloy buttons with separate embedded drawn copper-alloy wire shanks

Nos 89-90 are possibly *c.* late 13th – *c.* mid-14th century. Identical buttons are recorded from *c.*14th-century contexts of the River Thames foreshore, London, however, they are similar to 17th-century buttons of the same type, therefore are possibly intrusive. These buttons are often high-tin content or have a polished tin coating, both of which may create an appearance of silver. Examples of silver possibly exist. No. 91 was found in an archaeological deposit together with other artefacts attributed to the 14th century. No. 92 is from a late medieval deposit, although it is perhaps intrusive and may be as late as the 16th century. Unless otherwise stated, decoration is moulded-in-relief.

89. Shallow biconvex; solid; high-tin content; undecorated; a circular-section simple looped shank retaining a separate circular-section oval iron link and a separate fragment of drawn circular-section copper-alloy wire link; D 10.5mm; shank L 6mm; linkage L 6mm. *South Somerset.*

27

90. Discoidal; solid; conical front; flat back; high-tin content; undecorated; a circular-section simple looped shank; D 11mm; shank L 4.5mm. *South Somerset.*

91. Biconvex; solid; a collet that formerly may have held a stone; an octofoil and a mould line on the back; a very long circular-section simple looped shank; D 12mm; shank L 11mm. Possibly a pendant. *Billingsgate, London.* After Egan and Pritchard 1991, fig 179 2nd from bottom.

92. Biconvex; solid; undecorated apart from a central pellet; a very long circular-section simple looped shank. *River Thames foreshore, London.*

Fig 6. Monumental effigy of Joan Burwaschs, Lady Mohun, 14th century, in the Lady Chapel, the undercroft of Canterbury Cathedral, Kent. Note ornate discoidal mounts, often mistaken for buttons, on the overtunic. After Stothard 1817.

28

Composite three- or five-piece die-stamped sheet copper-alloy buttons with separate drawn copper-alloy wire shanks

Unless otherwise stated, all are three-piece. Nos 93-94 are from a secure *c.*1270 – *c.*1350 archaeological context; no. 95 was recovered from a late medieval deposit and no. 96 a secure 13th- 14th-century deposit. No. 97 came from an unstratified context, although found in spoil attributable to *c.*1270 – *c.*1380; this type, however, continued into at least the 17th century. Components are soldered together. Unless otherwise stated, decoration is punched.

93. Biconvex; hollow; undecorated; a (?) circular-, oval- or hemispherical-section simple looped shank; incomplete, part of front broken off; D 11mm; shank L 4mm. *Swan Lane, Upper Thames Street, London.* MoL, SWA81 2740 (2065). After Egan and Pritchard 1991, no. 1398.

94. Biconvex; hollow; undecorated; a circular-section simple looped shank; D 13mm; shank L 7mm. *Trig Lane, Upper Thames Street, London.* MoL, TL74 2121. After Egan and Pritchard 1991, no. 1404.

95. Plano-convex; hollow; undecorated; a (?) circular-, oval- or hemispherical-section simple looped shank; D 8mm; shank L 4mm. *Swan Lane, Upper Thames Street, London.* MoL, SWA81 685 (2072). After Egan and Pritchard 1991, no. 1402.

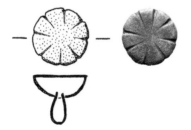

96. Plano-convex; hollow; eight radiating grooves that create a cusped edge; a circular-section simple looped shank; D 15mm; shank L 7mm. *River Thames foreshore, London.*

29

97. Plano-convex, hollow; five-piece; front has a raised band forming a rim, inside the rim is a band of rectangles each enclosing a quatrefoil (see enlargement); a (?) oval-section simple looped shank; incomplete, part of front broken off; D 22mm; shank L 6mm. *Swan Lane, Upper Thames Street, London.* MoL, SWA81 1808 (2081). After Egan and Pritchard 1991, no. 1403.

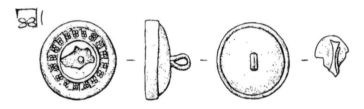

Composite three-piece die-stamped sheet copper-alloy button with separate soldered sheet copper-alloy shank
From a *c.*14th-century deposit. Components are soldered together. Decoration is engraved and openwork. Detached fronts are easily mistaken for 14th-century thimble crowns.

98. Biconvex; hollow; a voided saltire, each arm with a band of drilled openwork; two quadrants have a trefoil of drilled openwork, while the other two are filled with pits; within a voided hatched linear band; the back has similar openwork, although their irregularity implies they are gouged; a rectangular-section simple looped shank; D 15mm; shank L 5.5mm. *River Thames foreshore, London.* Another example of this rare type of button is published in *DA*, fig 179, bottom.

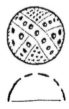

99. Front of a composite biconvex; hollow button similar to no. 98; D16mm. *Cambridgeshire.* UKDFD 144.

Composite three-piece die-stamped sheet copper-alloy buttons with separate soldered drawn copper-alloy wire shanks

All are from a *c.*15th-century context. Components are soldered together. Unless otherwise stated, decoration is die-stamped.

100. Biconvex; squashed; hollow; (?) a double multi-petalled rose; a circular-section simple looped shank; (note anaerobic gilding). *River Thames foreshore, London.*

101. Biconvex; squashed; hollow; indeterminate decoration; a circular-section simple looped shank; (note anaerobic gilding). *River Thames foreshore, London.*

102. Biconvex; squashed; hollow; a central (?) annulet; a circular-section simple looped shank; (note anaerobic gilding). *River Thames foreshore, London.*

103. Biconvex; hollow; a central pellet within an eight-spoked wheel, each spoke terminating in a pellet, within a beaded border; a circular-section simple looped shank; (note anaerobic gilding). *River Thames foreshore, London.*

104. Biconvex; hollow; undecorated; a circular-section simple looped shank; (note anaerobic gilding). *River Thames foreshore, London.*

105. Biconvex; hollow; undecorated; a circular-section simple looped shank; (note anaerobic gilding). *River Thames foreshore, London.*

Composite three-piece die-stamped sheet copper-alloy and sheet iron button with separate soldered drawn copper-alloy wire shank

From a *c.*15th-century context. Components are soldered together. Decoration is die-stamped.

106. Biconvex; squashed; hollow; an annulet within a raised circle; frontal corrosion is possibly decomposed paint; iron back; a circular-section simple looped shank; (note anaerobic gilding). *River Thames foreshore, London.*

Fig 7. Monumental effigy of William of Windsor and Blanche de la Tour, children of Edward III, *c.*14th century, in the Chapel of St Edmund, Westminster Abbey, London. Note discoidal buttons on tunic, and sleeve of the dress. After Stothard 1817.

Composite three-piece die-stamped sheet silver button with separate soldered drawn silver wire shank

Possibly *c*.13th – *c*.14th century, although this type continued into at least the 17th century. Components are soldered together.

107. Spherical; hollow; undecorated; a hemispherical-section simple looped shank; D 10mm; shank L 5mm. *South-West Wiltshire.*

Composite three-piece die-stamped sheet copper-alloy buttons with lead/tin alloy-filled backs and separate embedded drawn copper-alloy wire shanks

Both are from a *c*.15th-century context. Whether the shanks are attached with solder or embedded in the lead/tin alloy is uncertain; possibly both methods are used. Also recorded with drawn iron wire shanks. Unless otherwise stated, decoration is die-stamped.

108. Discoidal; solid; shallow convex front; flat back; undecorated; a circular-section simple looped shank; (note anaerobic gilding). *River Thames foreshore, London.*

109. Discoidal; solid; shallow convex front; flat back; a central boss with a circular pit within a raised circle; a circular-section simple looped shank; (note anaerobic gilding). *River Thames foreshore, London.*

Fig 8. Figures on the tomb of Sir Roger de Kerdeston (died 1337), representing relatives of the deceased, in Reepham Church, Norfolk. Note discoidal buttons on tunic, dress and cape. After Stothard 1817.

7: Post-Medieval Buttons

With regard to many post-medieval buttons, establishing where a certain type was worn on dress, the above observations concerning medieval buttons equally apply.

Cast one-piece copper-alloy buttons with integral drilled shanks

Nos 110, 114-27, 132 are from a *c*.16th-century context, although they may be as early as late 15th, perhaps as are nos 128-31, 133. Nos 111-13 are *c*.16th century. Nos 134-46 are *c*.17th century, although some probably continued into the early 18th. No. 147 is perhaps late 17th early 18th century. An extremely long shank is a typical feature found on some buttons of this type. Unless otherwise stated, decoration is moulded-in-relief. Datable evidence from the River Thames foreshore, London indicates that convex foil-backed glass-fronted metal buttons like no. 147, which may be an incomplete cuff-link, are perhaps attributable to the late 17th century or early 18th. Foil-backed buttons are difficult to differentiate from reverse-painted buttons of the 19th and 20th centuries; the style of shank is perhaps the best clue.

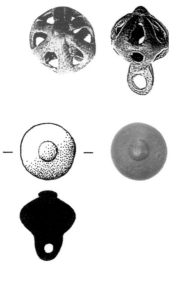

110. Biconvex; hollow; a central boss within a six-spoked wheel and ovoid openwork; the back similar but has oval openwork; a rectangular-section oval shank; (note anaerobic gilding). *River Thames foreshore, London.*

111. Biconvex; solid; undecorated apart from a central pellet; a rectangular-section rounded shank; D 15.5mm; shank L 8mm. *South Somerset.*

112. Spherical; solid; undecorated apart from a central elongated pellet, narrower at the bottom; incomplete, a rectangular-section rounded shank with the end broken off; D 11.5mm; shank L 4mm. *East Devon.*

35

113. Spherical; solid; undecorated; a rectangular-section rounded shank; D 12.5mm; shank L 5mm. *South Somerset.*

114. Discoidal; solid; convex front; flat back; a four-pointed star with a central pellet and a border of alternate voided chevrons with transverse lines and inverted voided U-shapes; a rectangular-section sub-rectangular shank with a rounded end; (note anaerobic gilding). *River Thames foreshore, London.*

115. Discoidal; solid; convex front; flat back; a central boss within a multi-spoked wheel; incomplete; a rectangular-section possible sub-rectangular shank with the end broken off; (note anaerobic gilding). *River Thames foreshore, London.*

116. Discoidal; solid; convex front; flat back; a central pellet within a quatrefoil; a rectangular-section sub-rectangular shank with a rounded end; (note anaerobic gilding). *River Thames foreshore, London.*

117. Discoidal; solid; convex front; flat back; undecorated apart from a central pellet; a rectangular-section sub-rectangular shank with a rounded end; (note anaerobic gilding). *River Thames foreshore, London.*

118. Sub-discoidal; solid; convex front; flat back; a central pellet bordered with six pellets; a rectangular-section roughly lozenge-shaped shank with a rounded end; (note anaerobic gilding). *River Thames foreshore, London.*

119. Discoidal; solid; convex front; flat back; a central pellet and a multifoil within a circle; a rectangular-section sub-rectangular shank with a rounded end; (note anaerobic gilding). *River Thames foreshore, London.*

120. Discoidal; solid; convex front; flat back; a five-spoked wheel; five sub-triangular panels each bearing a leaf on a black paint or niello field; a rectangular-section sub-rectangular shank with a rounded end. *River Thames foreshore, London.*

121. Discoidal; solid; convex front; flat back; a much abraded central (?) sexfoil bordered with radiating lines; a rectangular-section sub-rectangular shank with a rounded end; (note anaerobic gilding). *River Thames foreshore, London.*

122. Discoidal; solid; convex front; flat back; undecorated; a lateral border; white-metal coated overall; a rectangular-section sub-rectangular shank with a rounded end; (note sporadic anaerobic gilding). *River Thames foreshore, London.*

123. Discoidal; solid; convex front; flat back; a central boss surmounted by a pellet within a multifoil; a lateral rim; a rectangular-section sub-lozenge-shaped shank; (note anaerobic gilding). *River Thames foreshore, London.*

124. Discoidal; solid; convex front; flat back; a central pellet and an anticlockwise eight-bladed impeller within a cabled border; a rectangular-section sub-lozenge-shaped shank; (note anaerobic gilding). *River Thames foreshore, London.*

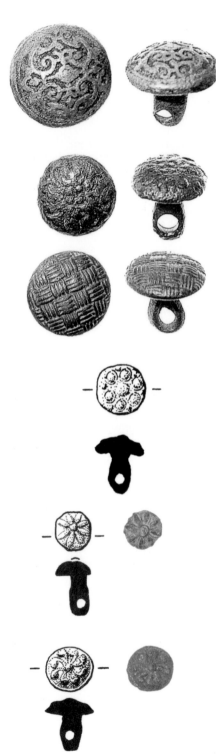

125. Discoidal; solid; convex front; flat back; conjoined curlicues within a circle; a rectangular-section sub-rectangular shank; (note anaerobic gilding). *River Thames foreshore, London.*

126. Discoidal; solid; convex front; flat back; a quatrefoil; a rectangular-section oval shank; (note anaerobic gilding). *River Thames foreshore, London.*

127. Discoidal; solid; convex front; flat back; basket weave; a rectangular-section oval shank with chamfered edges; (note anaerobic gilding). *River Thames foreshore, London.*

128. Discoidal; solid; convex front; flat back; a central pellet bordered with six pellets, each pellet within an annulet; bordered with small pellets; a rectangular-section oval long shank; D 14mm; shank L 14mm. *South Devon.* After Read 1995, no. 757.

129. Discoidal; solid; convex front; flat back; a central pellet and an octofoil; a rectangular-section sub-rectangular offset shank with a rounded end; D 10mm; shank L 11mm. *South-East Lincolnshire.*

130. Discoidal; solid; convex front; flat back; a central pellet and a clockwise eight-bladed impeller; a rectangular-section sub-lozenge-shaped shank with an angular end; D 13mm; shank L 10mm. *North-West Dorset.*

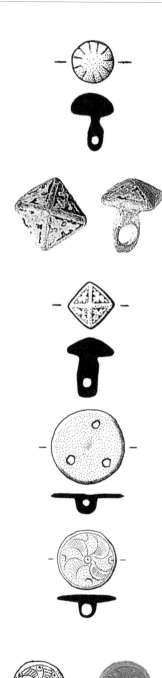

131. Discoidal; solid; convex front; flat back; bordered with radiating grooves; a rectangular-section sub-rectangular shank with a rounded end; D11.5mm; shank L 9.5mm. *Kent.* UKDFD 17383.

132. Lozenge-shaped; solid; pyramidal front; flat back; each face has a trefoil on a sporadic black paint or niello field; a rectangular-section oval shank; (note anaerobic gilding). *River Thames foreshore, London.*

133. Lozenge-shaped; solid; pyramidal front; flat back; decorated as no. 132, but shank is different; W 13mm; shank L 11mm. *South-West Wiltshire.*

134. Discoidal; flat; a central very shallow pit and three small holes; white-metal coated overall; a rectangular-section rounded shank; D 27mm; shank L 6mm. *South-East Lincolnshire.*

135. Shallow-biconvex; a central circle-and-dot motif and an engraved clockwise hatched four-bladed impeller, circle-and-dot motifs between the blades, within a circle; a rectangular-section rounded shank; D 24mm; shank L 6mm. *South Somerset.*

136. Discoidal; flat; a central circle-and-dot motif and an engraved clockwise hatched six-bladed impeller, circle-and-dot motifs between the blades, within a circle; a rectangular-section rounded shank; D 14mm; shank L 6mm. *South Somerset.*

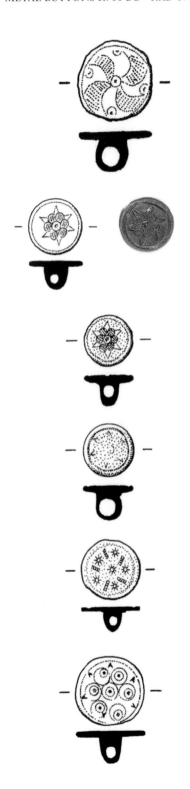

137. Discoidal; flat; a central circle-and-dot motif and an engraved clockwise hatched four-bladed impeller, circle-and-dot motifs between the blades within a circle; a rectangular-section rounded shank; D 16mm; shank L 6.5mm. *South Somerset.*

138. Discoidal; flat; an engraved sexfoil with pointed sepals within a circle; a rectangular-section rounded shank; D 15.5mm; shank L 6mm. *South Somerset.*

139. Discoidal; flat; decorated as no. 138, but shank is different; D 14mm; shank L 6mm. *South Somerset.*

140. Discoidal; flat; much abraded decoration as nos 138-39, but shank is different; D 14mm; shank L 7.5mm. *South Somerset.*

141. Discoidal; flat; engraved alternate eight-pointed stars and radiating rectangular hatched panels within a circle; a rectangular-section rounded shank; D 16mm; shank L 5mm. *South Somerset.*

142. Discoidal; flat; engraved circle-and-dot motifs, each within a circle, and small chevrons within a circle; resembles a sexfoil with pointed sepals; a rectangular-section rounded shank; D 19mm; shank L 7mm. *South Somerset.*

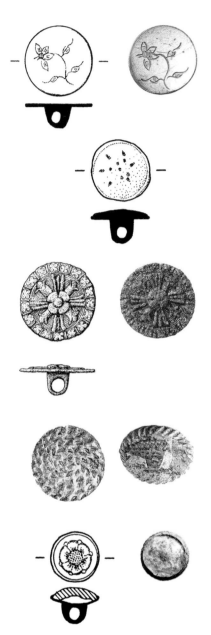

143. Discoidal; flat; an engraved three-leafed sprig and a quatrefoil; white-metal coated overall; a rectangular-section rounded shank; D 18mm; shank L 5.5mm. *South Somerset.*

144. Discoidal; solid; shallow convex front; flat back; engraved triangles and dots, resembles a sexfoil; a lateral rim; a rectangular-section rounded shank; D 16mm; shank L 6mm. *East Devon.*

145. Discoidal; flat; a central quatrefoil on a saltire of trefoil sepals and a cross with triple strand arms on a hatched field within a segmented border of small quatrefoils; a wedged-section trapezoid shank with a rounded end; D 23.5mm; shank L 7mm. *South Somerset.*

146. Discoidal; flat; imitative interwoven wire; a rectangular-section trapezoid shank; D 26mm; shank L 5mm. *Wiltshire.* WHM, acc. no. 26.1980.

147. Biconvex; a possible foil-backed clear glass dome; the inside of the back is slightly concave; the (?) foil decoration is a dark blue or black six-petalled rose with sepals on an off-white field; a wedged-section rounded shank; D 13.5mm; shank L 5mm. *South-East Dorset.* A similar object, the glass front of which is decorated with a cut, impressed or moulded multi-pointed star, is in WHM, acc. no. 1995.125.1.

41

Fig 9. Robert Dudley, Earl of Leicester 1532?–88, anonymous artist, English, *c.*1570. Note elaborate hemispherical buttons on doublet; in Montacute House, Somerset, NT.

Fig 10. Sir Thomas Cavandish 1560–92, buccaneer and the second Englishman to circumnavigate the world (1586–88), English, late 16th century. Note elaborate discoidal possible buttons on cape and possible hemispherical buttons on doublet; in Montacute House, Somerset, NT.

Fig 11. The early 17th-century almost unique painted (anonymous artist), ceiling of Muchelney Church, Somerset. Depicting ten angels, some very effeminate, garbed in Tudor costumes, eight of which have plain discoidal buttons on their doublets.

Cast one-piece copper-alloy buttons or cloak-fasteners with integral drilled shanks
Engraved and/or punched ornate floral designs are probably the most ubiquitous decoration on these buttons, followed by geometric and figurative. Moulded-in-relief decoration, sometimes very high, is known, but scarce, and while only three champlevé enamelled examples were recorded in the first edition, this number has now increased to five, although it is possible that some of the very high-relief examples were originally enamelled. For these buttons catalogued herein, unless otherwise stated, decoration is engraved or punched, or a combination of both. The Commonwealth arms depicted on no. 191 provides a safe mid-17th-century attribution. Nos 185-88 reputedly bear busts of King Charles II and Queen Catherine of Braganza respectively, presumably to commemorate their marriage in 1662; if true, thereby providing a probable secure date. River Thames foreshore, London, 17th-century contexts or datable archaeological spoil heaps in London have produced many of these buttons, including the aforementioned Commonwealth arms example, for members of the Society of Thames Mudlarks. The datable evidence indicates that these buttons became fashionable about the middle of the 17th century and they perhaps extended well into the 18th as suggested by a 1730 engraving of Viscount Cobham which shows possible enamelled buttons, perhaps of this type, on his coat front and cuffs (Fig. 12). Of course, one cannot be certain, for the backs are not visible. Frustratingly, this form of button, enamelled or otherwise, is not found on any surviving 17th- or 18th-century costume in the V&A Museum, Bath Museum of Costume or Platt Hall.

Interestingly, among the London finds are paired matching buttons of this form, joined shank to shank with either a drawn copper-alloy or iron wire linkage, which implies they were also used as cloak-fasteners (see no. 151). Curiously, paired buttons remain unrecorded from inland sites in Britain; however, when found, no. 196 had three rusted chain-links attached, of which one survives, suggesting use as a cloak-clasp. Inland, single specimens of this button turn up regularly and are recorded with drawn copper-alloy wire clips attached through their shanks (see no. 169), while others retain rust in their shank-holes, thereby implying they also are perhaps cloak-fasteners. It seems therefore that these items possibly performed a dual role as both true buttons and cloak-fasteners, perhaps worn en-suite.

148. Discoidal; solid; convex front; flat back; a sexfoil with annulet-headed sepals bordered with crescents; a rectangular-section trapezoid shank with a rounded end; D 27mm; shank L 5mm. *North-West Dorset.*

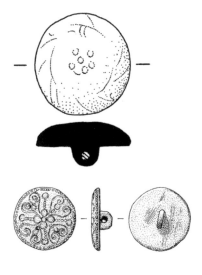

149. Discoidal; solid; convex front; flat back; much abraded; a cinquefoil formed from annulets within a circle; a hatched border; a rectangular-section trapezoid shank with a rounded end; rust in shank hole is probably from an iron clip, link or chain; D 27mm; shank L 5mm. *North Yorkshire.*

150. Discoidal; solid; convex front; flat back; a central quatrefoil with radiating spirals and bands of tiny and large annulets bordered with tiny annulets; a rectangular-section sub-rectangular shank; D 17.2mm; shank L 5.7mm. *South-East Dorset.* SSWM. ID. 2481.

151. Discoidal, a linked pair, each identical; solid; convex front; flat back; a quatrefoil; annulets resembling a bunch of grapes in each quadrant; bordered with annulets; a rectangular-section trapezoid shank with a rounded end; a drawn copper-alloy twisted wire link attached through each shank hole; D 26mm; shank L 5mm; link L 32mm. *River Thames foreshore, London.*

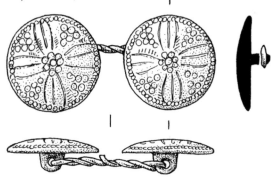

152. Discoidal; convex front; concave back; an ornate quatrefoil with sepals resembling bunches of grapes and two crescents in each quadrant; a rectangular-section trapezoid shank with a rounded end; D 26mm; shank L 7mm. *River Thames foreshore, London.*

45

153. Discoidal; convex front; concave back; a quatrefoil with fleuret-headed sepals bordered with crescents – highlighted with black paint; an offset rectangular-section trapezoid shank with a rounded end; D 26mm; shank L 8mm. *South-West Wiltshire.*

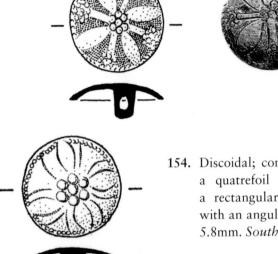

154. Discoidal; convex front; concave back; a quatrefoil bordered with annulets; a rectangular-section trapezoid shank with an angular end; D 24mm; shank L 5.8mm. *South Somerset.*

155. Discoidal; convex front; concave back; dots and annulets forming a quatrefoil with fleuret-headed sepals bordered with crescents; a rectangular-section trapezoid shank with a slightly rounded end; D 26mm; shank L 7mm. *East Devon.*

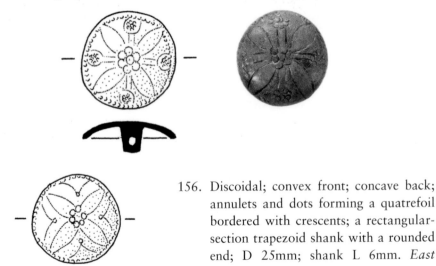

156. Discoidal; convex front; concave back; annulets and dots forming a quatrefoil bordered with crescents; a rectangular-section trapezoid shank with a rounded end; D 25mm; shank L 6mm. *East Devon.*

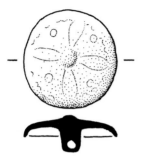

157. Discoidal; flattened convex front; concave back; a central moulded-in-relief pellet, a much abraded quatrefoil with an annulet in each quadrant bordered with crescents; a rectangular-section trapezoid shank with a slightly rounded end; D 27mm; shank L 6mm. *South-East Lincolnshire.*

158. Discoidal; convex front; concave back; dots forming a sexfoil on a field of dots bordered with crescents; incomplete, a rectangular-section trapezoid shank with a small piece broken off; D 25mm; shank L 9mm. *East Devon.*

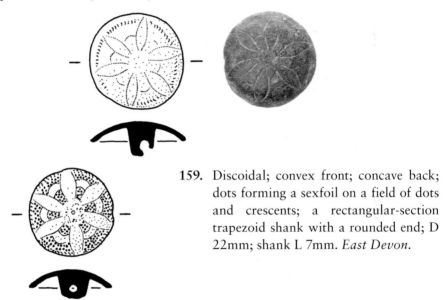

159. Discoidal; convex front; concave back; dots forming a sexfoil on a field of dots and crescents; a rectangular-section trapezoid shank with a rounded end; D 22mm; shank L 7mm. *East Devon.*

160. Discoidal; convex front; concave back; a quatrefoil with pointed sepals on a field of dots bordered with annulets; sporadic gilding on the front; a rectangular-section trapezoid shank with a rounded end; D 28mm; shank L 6mm. *Buckinghamshire.*

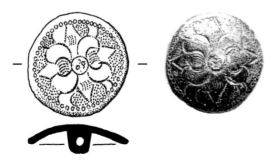

161. Discoidal; convex front; concave back; a cinquefoil with pointed sepals bordered with crescents; a slightly offset rectangular-section trapezoid shank with a partially rounded end; rust in the shank hole is probably from an iron link, clip or chain; D 27mm; shank L 6mm. *South Somerset.*

162. Discoidal; convex front; concave back; a quatrefoil on a field of dots bordered with crescents; a slightly offset rectangular-section trapezoid shank with a partially rounded end; D 25mm; shank L 6mm. *East Devon.*

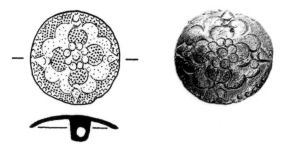

163. Discoidal; convex front; concave back; a cinquefoil within a concentric circle and a cinquefoil; bordered with tiny annulets; a rectangular-section trapezoid shank; D 25mm; shank L 7.5mm. *South-East Dorset.*

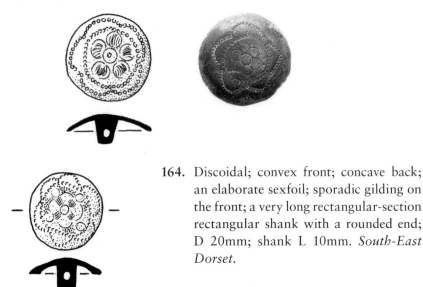

164. Discoidal; convex front; concave back; an elaborate sexfoil; sporadic gilding on the front; a very long rectangular-section rectangular shank with a rounded end; D 20mm; shank L 10mm. *South-East Dorset.*

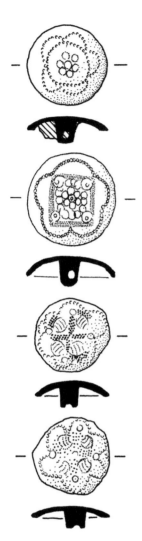

165. Discoidal; convex front; concave back; annulets forming a cinquefoil; a rectangular-section trapezoid shank with a slightly rounded end; rust on the back and in the shank hole is probably from an iron link, clip or chain; D 22mm; shank L 6mm. *South Somerset.*

166. Discoidal; convex front; concave back; annulets, crescents and dots forming a cinquefoil; a rectangular-section trapezoid shank with a rounded end; D 24mm; shank L 7mm. *Buckinghamshire.*

167. Sub-discoidal; convex front; concave back; dots and annulets forming a sexfoil bordered with crescents; incomplete, a rectangular-section rectangular shank with the end broken off; D 21mm; shank L 6mm. *South Somerset.*

168. Sub-discoidal; convex front; concave back; annulets and dots forming a sexfoil; incomplete, a rectangular-section trapezoid shank with the end broken off; D 21mm; shank remnant L 5mm. *East Devon.*

169. Discoidal; convex front; concave back; annulets, dots and crescents forming foliate, floral sprays and bunches of grapes; a rectangular-section trapezoid shank; a separate drawn copper-alloy wire clip is attached to the shank; D 27mm; shank L 8mm. *North-West Dorset.*

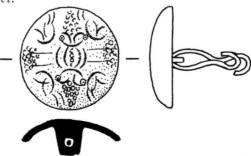

49

170. Discoidal; convex front; concave back; much abraded dots forming a (?) triple floral spray; incomplete, a rectangular-section trapezoid shank with a small piece broken off; D 27mm; shank L 7.5mm. *East Devon.*

171. Discoidal; convex front; concave back; much abraded dots forming a (?) triple floral spray bordered with dots; a rectangular-section trapezoid shank with a rounded end; D 25mm; shank L 6mm. *South Somerset.*

172. Discoidal; convex front; concave back; a spray of flowers and foliate; a rectangular-section trapezoid shank with a rounded end; D 26mm; shank L 7mm. *South Somerset.*

173. Discoidal; convex front; concave back; annulets forming a saltire with a crescent in each quadrant; a rectangular-section trapezoid shank with a slightly rounded end; D 24mm; shank L 5mm. *East Devon.*

174. Discoidal; convex front; concave back; a voided saltire with a central circle-and-dot motif, crescents and annulets resembling bunches of grapes in each quadrant; incomplete, an offset rectangular-section possible trapezoid shank with the end broken off; D 23.5mm; shank remnant L 5mm. *South Somerset.*

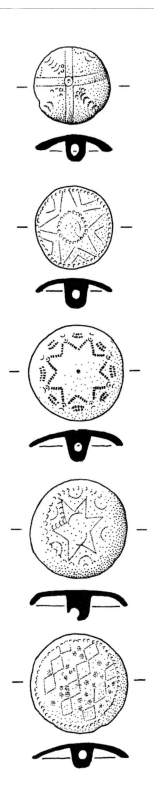

175. Discoidal; convex front; concave back; a voided saltire with a central circle-and-dot motif, annulets and crescents in each quadrant form a saltire; a rectangular-section rectangular shank with a rounded end; D 22mm; shank L 6mm. *South Somerset.*

176. Discoidal; convex front; concave back; annulets and dots forming a five-pointed star with a central circle, a chevron in each quadrant, bordered with annulets; a rectangular-section trapezoid shank with a rounded end; D 22.5mm; shank L 6mm. *East Devon.*

177. Discoidal; convex front; concave back; a central dot and an eight-pointed star formed from dashes, dots and crescents; bordered with alternate crescents and bands of dots; a rectangular-section trapezoid shank with a rounded end; D 26mm; shank L 7mm. *South-East Dorset.*

178. Discoidal; flattened convex front; concave back; a six-pointed star within a circle of small crescents, dots and larger crescents in the angles; incomplete, a rectangular-section trapezoid shank with part of the rounded end broken off; D 27mm; shank L 7mm. *South-West Wiltshire.*

179. Discoidal; convex front; concave back; three bands of contiguous lozenges on a field of dots bordered with annulets; a rectangular-section trapezoid shank with a slightly rounded end; D 27mm; shank L 6mm. *East Devon.*

180. Discoidal; convex front; concave back; three bands of contiguous lozenges on a field of dots and annulets bordered with annulets; a rectangular-section trapezoid shank with a rounded end; D 29mm; shank L 6mm. *River Thames foreshore, London.*

181. Discoidal; convex front; concave back; four bands of contiguous lozenges on a field of dots bordered with crescents and annulets; a rectangular-section trapezoid shank with a rounded end; D 28mm; shank L 8mm. *East Devon.*

182. Discoidal; convex front; concave back; a crowned facing mask of a king with small sexfoil and curlicues in the field bordered with annulets; a rectangular-section trapezoid shank with a rounded end; D 28mm; shank L 9mm. *River Thames foreshore, London.*

183. Discoidal; convex front; concave back; a crowned facing mask of a king with crescents and lines in the field bordered with annulets; a rectangular-section trapezoid shank; D 23mm; shank L 7mm. *Surrey.*

184. Discoidal; convex front; concave back; a very high moulded-in-relief facing bust of a cavalier; a rectangular-section trapezoid shank with a rounded end; incomplete, edge ragged and bent; D 29mm; shank L *c.*9mm. *South Devon*. After Read 1995, no. 980.

185. Discoidal; convex front; concave back; a moulded-in-relief profile right bust of (?) King Charles II; a rectangular-section rectangular shank with a rounded end; D 28mm; shank L *c.*9mm. *North Wiltshire*. WHM, acc. no. DM 516. After Giltsoff and Robinson [no.3].

186. Discoidal; convex front; concave back; comparable with no. 185; a moulded-in-relief profile right bust of (?) King Charles II; a rectangular-section trapezoid shank with a rounded end; D 25mm; shank L *c.*5mm. *East Devon*.

187. Discoidal; convex front; concave back; a moulded-in-relief profile left bust of (?) Queen Catherine of Braganza; sporadic white-metal coating overall; a rectangular-section rectangular shank with a rounded end; D *c.*27mm; shank L *c.*9mm. *Wiltshire*. WHM, acc. no. 1991.36.

188. Discoidal; convex front; concave back; comparable with no. 187; a moulded-in-relief profile left bust of (?) Queen Catherine of Braganza; a rectangular-section trapezoid shank with a rounded end; D 24.5mm; shank L *c.*7mm. *South Somerset.*

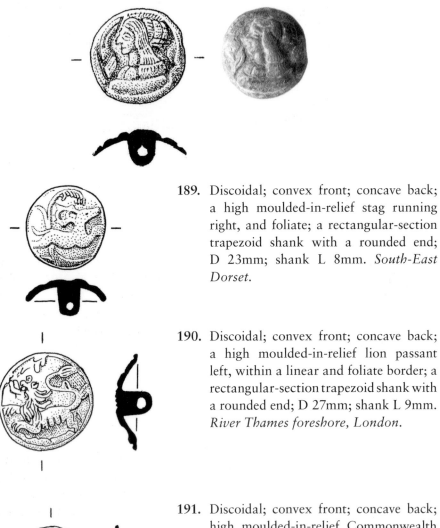

189. Discoidal; convex front; concave back; a high moulded-in-relief stag running right, and foliate; a rectangular-section trapezoid shank with a rounded end; D 23mm; shank L 8mm. *South-East Dorset.*

190. Discoidal; convex front; concave back; a high moulded-in-relief lion passant left, within a linear and foliate border; a rectangular-section trapezoid shank with a rounded end; D 27mm; shank L 9mm. *River Thames foreshore, London.*

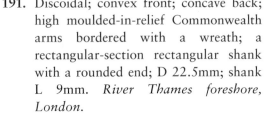

191. Discoidal; convex front; concave back; high moulded-in-relief Commonwealth arms bordered with a wreath; a rectangular-section rectangular shank with a rounded end; D 22.5mm; shank L 9mm. *River Thames foreshore, London.*

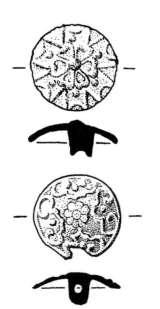

192. Discoidal; convex front; concave back; a high moulded-in-relief central pellet within a double sexfoil; incomplete, a rectangular-section trapezoid shank with the end broken off; D 24mm. *South Devon*. After Read 1995, no. 979.

193. Discoidal; convex front; convex back; a high-relief octofoil bordered with curlicues and pellets; incomplete, section of rim broken off; a rectangular-section trapezoid shank; D 24mm; shank L 7.2mm. *South Somerset*.

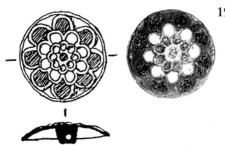

194. Discoidal; convex front; concave back; a blue-and-white champlevé enamelled triple octofoil within a circle; a rectangular-section trapezoid shank; D 24mm; shank L *c*.6mm. *South Devon*. After Read 1995, no. 982.

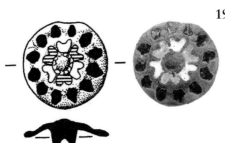

195. Discoidal; convex front; concave back; a light-blue and white champlevé enamel sexfoil bordered with dark-blue champlevé enamel asymmetrical shapes; a rectangular-section trapezoid shank; D 23mm; shank L 9mm. *South Somerset*.

196. Discoidal; much corroded; convex front; convex back; what appears to be a light-blue and white champlevé enamel sexfoil bordered with light- and dark-blue champlevé enamel semicircles; a rectangular-section trapezoid shank retaining a fragment of chain-link, now solid rust (when found, three links survived but two have since disintegrated); D 21.2mm; shank L *c.*3mm; chain-link L 18mm. *South Somerset.*

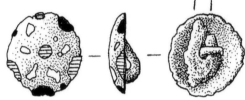

197. Discoidal; convex front; concave back; a horse and rider right, the horseman holding a (?) musket in his right hand, a dark-blue and white champlevé enamel field; a rectangular-section shank with the end broken off; D 15.3mm; shank remnant L 3mm. *East Devon.*

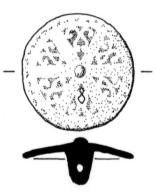

198. Discoidal; convex front; concave back; a much abraded moulded-in-relief six-spoked wheel with a central pellet, S-spirals in each angle, a cabled border; white-metal coated overall; one casting-hole; a rectangular-section trapezoid shank with a rounded end; D 31mm; shank L 10mm. *South Somerset.*

199. Discoidal; convex front; concave back; a moulded-in-relief six-spoked wheel with a central pellet, S-spirals in each angle, a voided cabled border; white-metal coated overall; several casting-holes; a rectangular-section trapezoid shank with a rounded end; D 32mm; shank L 9mm. *South Somerset.*

200. Discoidal; convex front; concave back; a moulded-in-relief six-spoked wheel with a central pellet; an S-spiral and openwork in each angle, within a knurled border; white-metal coated overall; a rectangular-section trapezoid shank with a rounded end; D 31.5mm; shank L c.9mm. *North Dorset.*

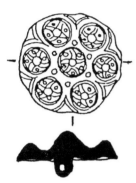

201. Discoidal; solid; moulded-in-relief; a sexfoil formed from a seven bosses, six tiny pellets surround the central boss, each boss has a quatrefoil with a central pellet and a tiny pellet in the quadrants; flat back; a rectangular-section trapezoid shank with a rounded end; D 28mm; shank L *c.*4mm. *South Devon.* After Read 1995, no. 981.

202. Hexagonal; convex front with faceted sides and flat top; concave back; dots forming a voided hexagon within a voided six-spoked wheel, annulets and curlicues in the angles, four annulets within a circle on the top; incomplete, a probable rectangular-section trapezoid shank broken off; W 25.5mm. *South Devon.* After Read 1995, no. 983.

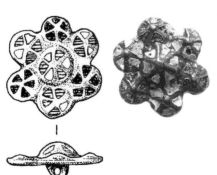

203. Sexfoil; a central boss; six peripheral bosses on a lateral rim; white, light- and dark blue champlevé enamel forming a central sexfoil and a smaller sexfoil on each lobe, a central pellet on each sexfoil; a rectangular-section rounded shank; D 27mm; shank L 4.5mm. *East Devon.*

Cast one-piece silver button or cloak-fastener with integral drilled shank
The style is typical mid- to late 17th century, however, the moulded-in-relief device is probably considerably later, therefore an 18th or 19th century attribution is likely. It is included here to highlight the pitfalls of dating buttons recovered from unstratified contexts.

204. Discoidal; convex front; concave back; an unidentifiable armorial device; a rectangular-section trapezoid shank with a rounded end; D 22mm; shank L 9mm. *East Devon.*

Fig 12. Viscount Cobham, redrawn from an engraving by J. Faber after G Kneller, 1730. Note discoidal convex possible enamelled buttons down the front and on the sleeves of the coat.

Fig 13. A stumpwork mirror depicting King Charles II (note double row of discoidal buttons on his chest) and Queen Catherine of Braganza, *c*.1670. Reputedly made in Amesbury, Wiltshire. In SSWM.

Fig 14. Part of the stumpwork mirror detail enlarged.

Cast one-piece copper-alloy buttons with integral undrilled shanks
Nos 205–13 are possibly late 15th – 16th century. Nos 214-21, 230-34 came from a *c.*16th – *c.*17th-century context. Nos 222-29 are *c.*16th century into the 17th. An extremely long shank is a typical feature found on some buttons of this type. Unless otherwise stated, decoration is moulded-in-relief.

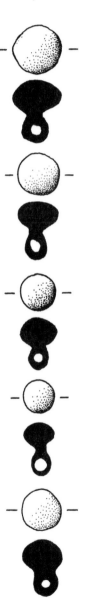

205. Biconvex; solid; undecorated; a circular-section short stem and a circular-section loop; D 13mm; shank L 7mm. *South-East Lincolnshire.*

206. Biconvex; solid; undecorated; a circular-section short stem and a circular-section loop; D 10mm; shank L 7mm. *North-West Dorset.*

207. Biconvex; solid; undecorated; a circular-section simple looped shank; D 10mm; shank L 6mm. *South-West Wiltshire.*

208. Biconvex; solid; undecorated; a circular-section short stem and a circular-section loop; D 8.5mm; shank L 7mm. *North-West Dorset.*

209. Biconvex; solid; undecorated; a circular-section simple looped shank; D 11mm; shank L 6mm. *East Devon.*

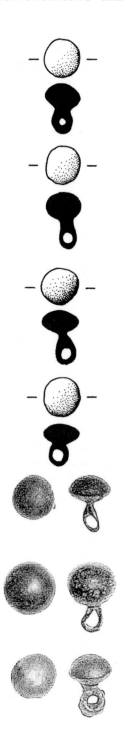

210. Biconvex; solid; undecorated; a mould line on the back; a circular-section simple looped shank; D 10mm; shank L 6mm. *East Devon.*

211. Biconvex; solid; undecorated; a circular-section medium length stem and a circular-section loop; D 10mm; shank L 7.5mm. *East Devon.*

212. Biconvex; solid; undecorated; a circular-section medium length stem and a circular-section loop; D 11mm; shank L 9mm. *South-West Wiltshire.*

213. Biconvex; solid; undecorated; a circular-section simple looped shank; D 10mm; shank L 6mm. *South-West Wiltshire.*

214. Biconvex; solid; undecorated; a circular-section short stem and a circular-section loop; (note anaerobic gilding). *River Thames foreshore, London.*

215. Biconvex; solid; undecorated; a circular-section short stem and a circular-section loop; (note anaerobic gilding). *River Thames foreshore, London.*

216. Biconvex; solid; undecorated; a mould line around the shank which has a casting sprue; a circular-section short stem and a circular-section loop; (note anaerobic gilding). *River Thames foreshore, London.*

217. Biconvex; solid; undecorated; a circular-section short stem and a circular-section loop; (note anaerobic gilding). *River Thames foreshore, London.*

218. Biconvex; solid; undecorated; a circular-section short stem and a circular-section loop; (note anaerobic gilding). *River Thames foreshore, London.*

219. Biconvex; solid; undecorated apart from a central pellet; a circular-section short stem and a circular-section loop; (note anaerobic gilding). *River Thames foreshore, London.*

220. Biconvex; solid; undecorated apart from a central pellet; a circular-section short stem and a circular-section loop; (note anaerobic gilding). *River Thames foreshore, London.*

221. Biconvex; solid; undecorated apart from a central pellet; a circular-section short stem and a circular-section loop. *River Thames foreshore, London*

222. Biconvex; solid; undecorated apart from a central pellet; incomplete, a circular-section medium length stem, a circular-section simple loop with part broken off; D 9mm; shank L *c.*7mm. *South-East Lincolnshire.*

223. Biconvex; solid; undecorated apart from a central pellet; incomplete, a circular-section medium length stem, a circular-section simple loop with part broken off; D 11mm; shank L 9mm. *South-East Lincolnshire.*

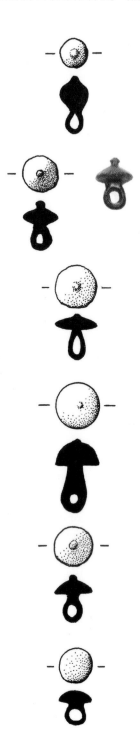

224. Biconvex; solid; undecorated apart from a central pellet; sporadic black paint overall; a circular-section simple looped shank; D 8mm; shank L 6mm. *East Devon.*

225. Biconvex; solid; undecorated apart from a central pellet; a circular-section simple looped shank; D 10mm; shank L 7mm. *North-West Dorset.*

226. Biconvex; solid; undecorated apart from a central pellet; a circular-section simple looped shank; D 12mm; shank L 8mm. *South-West Wiltshire.*

227. Discoidal; solid; convex front; flat back; undecorated apart from a central pellet; a circular-section long stem and a circular-section loop; D 12.5mm; shank L 13mm. *South-West Dorset.*

228. Discoidal; solid; convex front; flat back; undecorated apart from a central pellet; a circular-section short stem and a circular-section loop; D 11mm; shank L 7.5mm. *South-East Lincolnshire.*

229. Discoidal; solid; convex front; flat back; undecorated; a circular-section simple looped shank; D 9mm; shank L 6mm. *South Somerset.*

230. Discoidal; solid; convex front; flat back; an ornate four-petalled rose surrounded by curlicues within a beaded border; a circular-section very short stem and a circular-section loop; (note anaerobic gilding). *River Thames foreshore, London.*

231. Discoidal; solid; convex front; flat back; a triple five-petalled rose within a beaded border; a circular-section very short stem and a circular-section loop; (note anaerobic gilding). *River Thames foreshore, London.*

232. Discoidal; solid; convex front; flat back; a double five-petalled rose within a beaded border; a circular-section simple looped shank retaining a separate drawn copper-alloy wire clip identical to the eye-section of a Read Class A single blunt-hooked and eye clasp; (note anaerobic gilding). *River Thames foreshore, London.*

233. Discoidal; solid; convex front; flat back; engraved and punched annulets and foliate within a circle bordered with foliate; a circular-section simple looped shank; (note anaerobic gilding). *River Thames foreshore, London.*

234. Discoidal; convex front; concave back; an engraved and punched six-pointed star with punched crescents in the angles within a circle of annulets, a central large annulet enclosing smaller annulets within a sexfoil border of annulets; a circular-section simple looped shank retaining a fragment of drawn copper-alloy wire clip or link; (note anaerobic gilding). *River Thames foreshore, London.*

Cast one-piece lead/tin alloy buttons with integral drilled shanks

Due to their large size and heaviness, some of these buttons seem impractical for use as multiple fasteners on individual items of clothing, though one or two for fastening a cloak is feasible. Perhaps a more logical function was for securing bags or satchels. All are c.17th century, but perhaps continued into the 18th. Decoration is moulded-in-relief, though two have applied surface decoration.

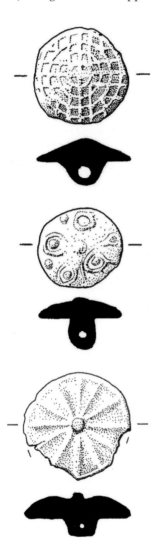

235. Discoidal; solid; convex front; flat back; chequering formed from four concentric circles overlaid by a triple-stranded cross, a chevron in each quadrant; a rectangular-section rounded shank; D 26mm; shank L 6mm. *North Devon.*

236. Discoidal; solid; convex front; flat back; a central pellet bordered with alternate circle-and-dot motifs and smaller pellets; resembles a quatrefoil; part of the decoration is miscast; a rectangular-section rounded shank; D 23mm; shank L 8.3mm. *South Wiltshire.*

237. Discoidal; solid; convex front; flat back; a multi-spoked wheel with a central pellet; a rectangular-section trapezoid shank; incomplete, edge ragged; D 30mm; shank L 5mm. *South Somerset.*

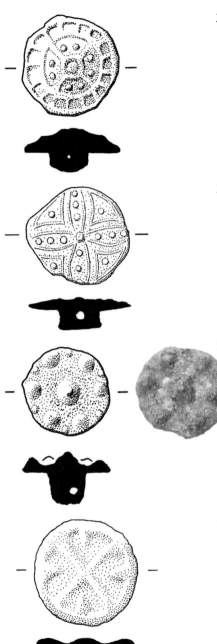

238. Discoidal; solid; convex front; flat back; three concentric circles and a saltire with two pellets in each quadrant, bordered with radiating ridges; a rectangular-section rectangular shank with a rounded end; D 26mm; shank L 5mm. *East Devon.*

239. Shallow biconvex; solid; a quatrefoil with a central pellet, three further pellets in each petal; each quadrant has a curved-sided triangle enclosing a pellet; a rectangular-section wide rectangular shank; D 28.5mm; shank L 5.5mm. *East Devon.*

240. Discoidal; flat; a central boss surmounted by a pellet bordered with alternate large bosses and small pellets; a rectangular-section trapezoid shank with a rounded end; D 25mm; shank L 8.5mm. *South Somerset.*

241. Discoidal; flat; a recessed quatrefoil; a rectangular-section trapezoid shank; D 29mm; shank L 9mm. *South Somerset.*

67

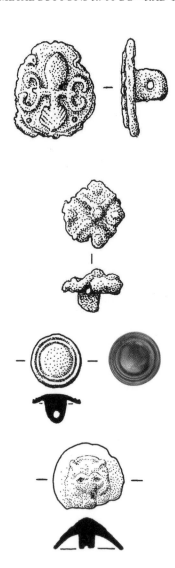

242. Discoidal; flat; a stylised fleur-de-lis – representing the purity of the Virgin Mary – and two tiny pellets at one end, within a circle; incomplete; edge ragged; a rectangular-section rectangular shank with a rounded end; estimated D 30mm; shank L 9mm. *North Dorset.*

243. Discoidal; flat; a saltire with a central pellet; a pellet in each quadrant; incomplete, edge ragged; the shank appears to have part of the end broken off and has been drilled close to the back; estimated D 20mm; shank remnant L 5.5mm. *Gloucestershire.*

244. Discoidal; convex front; concave back; a border of two concentric circles; a rectangular-section rounded shank; D 14mm; shank L 7mm. *South-West Wiltshire.*

245. Discoidal; convex front; concave back; a facing mask of a lion with protruding tongue; incomplete, edge ragged and a rectangular-section shank with the end broken off; D 20mm; shank remnant L 4.5mm. *South Somerset.*

246. Discoidal; convex front; concave back; white-metal coated overall; a right-handed spiral imitating a coiled chain; a rectangular-section rectangular shank; D 22mm; shank L 6.2mm. *County Durham.* UKDFD 5808.

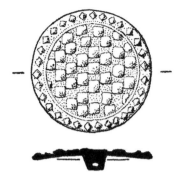

247. Discoidal; shallow convex front; shallow concave back; a chequered circle bordered with lozenge-shaped pellets; sporadic black paint on the front; a rectangular-section trapezium-shaped shank; D 33mm; shank L 4mm. *County Durham*. UKDFD 5576.

Cast one-piece lead/tin alloy buttons with integral undrilled shanks

The comments on use made in the immediately preceding large lead/tin alloy buttons equally apply here. An extremely long shank is a typical feature on some buttons of this type. Nos 257-61, 311, 314 are from secure 16th – 17th-century contexts. Nos 248-56, 262-80, 310, 312, 365 are *c.*16th century, possibly extending into the early 17th. Nos 281-302, 305-09, 313 are perhaps late 16th – 17th century. Nos 328-64 are *c.*17th century though some may have continued into the early 18th. No. 303 is 1603-25; no. 304 could be late 16th century, while no. 362 may commemorate James VI of Scotland 1603 accession as James I to the English throne. Profile bust or horse-and-rider depictions of King Charles I always face left on his coinage, therefore, if the profile bust on no. 327 is Charles I, a secure date of between 1625 and 1649 is probable. However, hammered coinage and the milled silver fourpence of King Charles II also have profile busts facing left, presenting the possibility that this button may have been made during his reign, 1660-85. Lead weights of both King Charles I and II are found stamped with a crowned C, however, some have the crowned initials C R. Charles II lead weights are sometimes stamped C II R surmounted by a crown. Therefore, such marks provide no help in establishing a firm date of manufacture for this button. Erring on the side of caution, an attribution of between 1625 and 1685 is safer. Unless otherwise stated, decoration is moulded-in-relief.

248. Biconvex; solid; undecorated; a circular-section short stem and a circular-section loop; D 10mm; shank L 8mm. *South-East Lincolnshire*.

249. Biconvex; solid; undecorated; a circular-section short stem and a circular-section loop; D 10mm; shank L 5.5mm. *South-East Lincolnshire*.

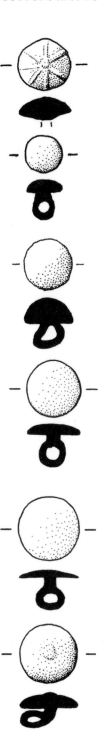

250. Biconvex; solid; a seven-spoked wheel and a central pellet; incomplete, shank broken off; D 15mm. *South Somerset.*

251. Discoidal; solid; convex front; flat back; undecorated; a circular-section very short stem and a circular-section loop; D 10mm; shank L 7mm. *South-East Lincolnshire.*

252. Discoidal; solid; convex front; flat back; undecorated; a circular-section simple looped shank; D 14mm; shank L 7mm. *South-West Wiltshire.*

253. Discoidal; solid; convex front; flat back; undecorated; a circular-section short stem and a circular-section loop; D 15.5mm; shank L 9mm. *South Somerset.*

254. Discoidal; solid; convex front; flat back; undecorated; a circular-section medium-length stem and a circular-section loop; D 16.5mm; shank L 9mm. *South Somerset.*

255. Discoidal; solid; convex front; flat back; undecorated apart from a central pellet; a circular-section short stem and a circular-section loop; D 16mm; shank L *c.*10mm. *North Dorset.*

256. Discoidal; solid; convex front; flat back; undecorated apart from a central pellet; a circular-section very wide simple looped shank; D 11mm; shank L 3mm. *South-West Wiltshire.*

257. Discoidal; solid; convex front; flat back; a central boss and a sexfoil on a hatched field; bordered with oval pellets; a mould line on the back and the shank; a circular-section simple looped shank. *River Thames foreshore, London.*

258. Discoidal; solid; convex front; flat back; a central pellet and a double five-petalled rose; a mould line on the back and the shank; a circular-section simple looped shank. *River Thames foreshore, London.*

259. Discoidal; solid; convex front; flat back; a central pellet and an octofoil within a multi-pitted border; a mould line on the back and the shank; a casting sprue on the shank; a circular-section simple looped shank. *River Thames foreshore, London.*

260. Discoidal; solid; convex front; flat back; an eight-spoked wheel, hatched in the angles; resembles a spider's web; a mould line on the back; a circular-section simple looped shank. *River Thames foreshore, London.*

261. Discoidal; solid; convex front; flat back; a voided saltire with a central pellet and a circle-and-dot motif in each quadrant; a circular-section short stem and a circular-section loop. *River Thames foreshore, London.*

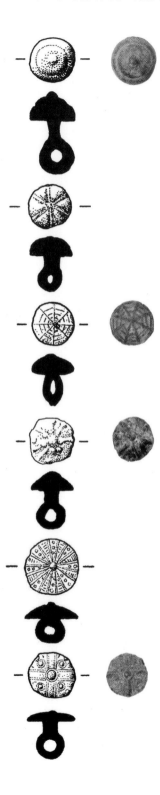

262. Discoidal; solid; convex front; flat back; a central boss surmounted by a pellet; a mould line on the back and the shank; a circular-section long stem and a circular-section loop; D 13mm; shank L 15mm. *South-West Wiltshire.*

263. Discoidal; solid; convex front; flat back; a central small pellet and a six-spoked wheel; a circular-section long stem and a circular-section loop; D 12mm; shank L 9mm. *South Somerset.*

264. Discoidal; solid; convex front; flat back; an eight-spoked wheel with a central pellet, hatched in the angles; resembles a spider's web; a circular-section simple looped shank; D 12mm; shank L 8.5mm. *South-East Lincolnshire.*

265. Discoidal; solid; convex front; flat back; a central pellet and an eight-spoked wheel with three smaller pellets in each angle; a circular-section very long stem and a circular-section loop; D 12mm; shank L 10mm. *South-West Wiltshire.*

266. Discoidal; solid; convex front; flat back; a central pellet and a multi-spoked wheel, the angles have alternate smaller pellets or hatching; a mould line on the back; a circular-section simple looped shank; D 15mm; shank L 7mm. *East Yorkshire.*

267. Discoidal; solid; convex front; flat back; a voided plain cross with a central pellet, a circle-and-dot motif and a tiny pellet in each quadrant; a circular-section long stem and a circular-section loop; D 14mm; shank L 10mm. *South-East Lincolnshire.*

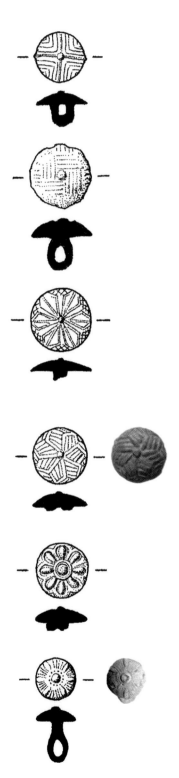

268. Discoidal; solid; convex front; flat back; a saltire and central pellet; four chevrons in each quadrant; a mould line on the back; a circular-section simple-looped shank; D 14mm; shank L 6.2mm. *County Durham.* UKDFD 5467.

269. Discoidal; solid; convex front; flat back; basket weave and a central pellet; a mould line on the back; a circular-section simple looped shank; D 17.5mm; shank L 9mm. *East Yorkshire.*

270. Discoidal; solid; convex front; flat back; a sexfoil and a central pellet within a cross-hatched border; a mould line on the back; incomplete, a circular-section short stem, circular-section loop broken off; D 17mm; shank remnant L 2mm. *Lancashire.*

271. Discoidal; solid; convex front; flat back; a hatched sexfoil and a central pellet; a mould line on the back; incomplete, a circular-section simple looped shank broken off; D 15mm; shank remnant L 1mm. *Cambridgeshire.* UKDFD 9041.

272. Discoidal; solid; convex front; flat back; an octofoil and a central pellet; a mould line on the back; incomplete, a circular-section simple looped shank broken off; D 14mm; shank remnant L 1.5mm. *South-West Wiltshire.*

273. Discoidal; solid; convex front; flat back; a sexfoil and a central pellet on a hatched field; a mould line on the back; a circular-section long stem and a circular-section simple loop; D 10.5mm; shank L 11mm. *County Durham.*

73

274. Discoidal; solid; convex front; flat back; a multifoil; a mould line on the back; a circular-section short stem and a circular-section loop; incomplete, edge ragged; D 17mm; shank L 6mm. *East Devon.*

275. Discoidal; solid; convex front; flat back; a five-pointed star with a central annulet within an elaborate cinquefoil; a mould line on the back; incomplete, a circular-section stem, probable circular-section loop broken off; D 16mm; shank remnant L *c.*6mm. *South-East Dorset.*

276. Discoidal; solid; convex front; flat back; a central pellet and a sexfoil on a hatched field; incomplete, edge miscast and a circular-section short stem, circular-section loop broken off; D 17mm; shank remnant L 2.9mm. *East Devon.*

277. Discoidal; solid; convex front; flat back; a central pellet and an anticlockwise hatched six-bladed impeller; a circular-section simple looped shank; D 15mm. shank L 6.5mm. *North Yorkshire.*

278. Discoidal; solid; convex front; flat back; a central pellet bordered with seven pellets; a slight projection on the side is probably a casting sprue; incomplete, shank broken off; D 15mm; shank remnant L 1mm. *South Somerset.*

279. Discoidal; solid; shallow convex front; flat back; a central pellet and circle within a stylised cross moline, two pellets in each quadrant; a mould line on the back; incomplete, edge ragged and shank broken off; D 28.3mm; shank remnant L 1.8mm. *North Yorkshire.* UKDFD 8645.

280. Discoidal; solid; convex front; flat back; a six-spoked wheel with a central pellet, a pellet in each angle; incomplete, a slightly ragged edge and shank broken off; D 15mm. *South Somerset.*

281. Discoidal; solid; shallow convex front; flat back; a quatrefoil with a central pellet and a pellet in each angle within a circle; a wide border of alternate pellets and ovoids; a mould line on the back; incomplete, circular-section shank broken off; D 30mm; shank remnant L 2mm. *South-West Dorset.*

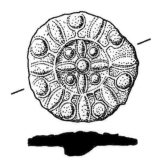

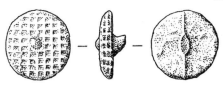

282. Discoidal; solid; convex front; flat back; a central pellet and cross-hatching; a mould line on the back; incomplete; an oval-section stem, loop broken off; D 19.5mm; shank remnant L 4mm. *South-West Wiltshire.*

283. Discoidal; solid; convex front; flat back; an eight-spoked wheel with a central pellet and a smaller pellet in each angle; an offset circular-section triangular shank; D 29mm; shank L 8mm. *East Devon.*

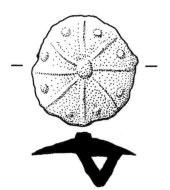

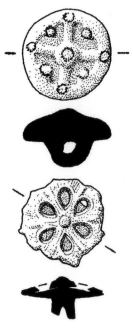

284. Discoidal; solid; convex front; flat back; a plain cross with a central boss and a pellet, each arm terminates in a pellet, each quadrant is recessed and has a pellet; a circular-section simple looped shank; D 24.5mm; shank L 7mm. *Lancashire.*

285. Discoidal; hollow; convex front; flat back; an openwork sexfoil with a central pellet; incomplete, a ragged edge and a circular-section short stem, circular-section loop broken off; D 23mm; shank remnant L 5.8mm. *Staffordshire.* UKDFD 11441.

286. Discoidal; solid; convex front; flat back; an eight-spoked wheel with a central pellet, each spoke has two longitudinal grooves; recessed between the spokes; a beaded border; a mould line on the back; a circular-section simple looped shank; D 30; shank L uncertain. *Lincolnshire.* UKDFD 17198.

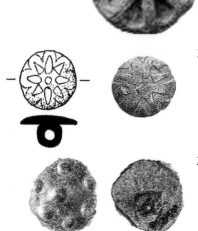

287. Discoidal; solid; convex front; flat back; a double octofoil and a central annulet; a mould line on the back; a circular-section simple looped shank; D 16mm; shank L 6mm. *South Somerset.*

288. Discoidal; solid; umbonate front; flat back; a central pellet bordered with six pellets; a circular-section simple looped shank; incomplete, edge ragged; D 24mm. *Wiltshire.* WHM, acc. no. 80.1980.

289. Discoidal; solid, convex front, flat back; undecorated apart from a central pellet; a circular-section long stem and a circular-section loop; D 12.5mm; shank L 13mm; comparable with no. 227. *South-West Dorset.*

290. Discoidal; solid; convex front; flat back; a multifoil within a sexfoil on a hatched field; a mould line on the back; incomplete, shank broken off; D 11mm; shank remnant L 2mm. *Lancashire.*

291. Discoidal; solid; convex front; flat back; a quatrefoil with two pellets and a crescent in each quadrant; a circular-section simple looped shank; D 25mm; shank L 7.5mm. *South-West Wiltshire.*

292. Discoidal; solid; convex front; flat back; an octofoil on a hatched field; a circular-section simple looped shank; D 15.8mm; shank L 8.5mm. *North Yorkshire.*

293. Discoidal; solid; convex front; flat back; an octofoil on a hatched field; incomplete; a circular-section stem, loop broken off; D 18mm; shank remnant L 2.5mm. *South Somerset.*

294. Discoidal; solid; convex front; flat back; two concentric circles bordered with radiating grooves, a circular-section long stem and a circular-section simple loop; D 20mm; shank L 10mm. *South Dorset.*

295. Discoidal; solid; convex front; flat back; annulets forming a plain cross within a voided linear plain cross, the inside edge of each arm is outlined with hatching; on a hatched field; a circular-section short stem and a circular-section simple loop; D 15mm; shank L 12.5mm. *East Yorkshire*. UKDFD 8383.

296. Discoidal; solid; convex front; flat back; basket weave within two concentric circles infilled with a band of dots; a circular-section long stem and a circular-section loop; D 6.9mm; shank L 7mm. *River Thames foreshore, London.*

297. Discoidal; solid; convex front; flat back; basket weave; a mould line on the back; incomplete, circular-section shank broken off; D 19.5mm. *East Yorkshire.*

298. Discoidal; solid; convex front; flat back; a saltire with a pellet in each quadrant; a rectangular-section rectangular shank with a rounded end; D 22mm; shank L 6mm. *East Devon.*

299. Discoidal; solid; convex front; flat back; much abraded geometric decoration; a mould line on the back; a curiosity, for the D-section simple looped shank has evidently been deliberately cut in two and bent over; the body, just above the rim, is pierced, presumably to take a cord; maximum D 27mm; shank L *c.*6mm. *East Devon.*

300. Discoidal; solid; convex front; flat back; a six-spoked wheel; a mould line on the back; a circular-section simple looped shank; D 21mm; shank L 6mm. *North Yorkshire.*

301. Discoidal; solid; convex front; flat back; a hatched sexfoil on a hatched field; a central pellet and a pellet in each petal, bordered with smaller pellets; a mould line on the back; a circular-section simple looped shank; D 20mm; shank L 6mm. *North Yorkshire.*

302. Discoidal; solid; convex front; flat back; a six-spoked wheel on a hatched field, bordered with pellets, vestige of a mould line on the back; incomplete, edge a little ragged and shank broken off; D 20mm. *North Yorkshire.*

303. Discoidal; solid; convex front; flat back; a profile bust left of either James VI of Scotland or James I of England, I left R right, within a cusped circle; a circular-section simple looped shank; D 28mm; shank L uncertain. *Lancashire.* PAS LANCUM-E363F2.

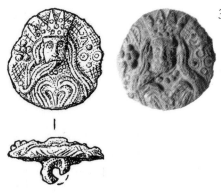

304. Discoidal; solid; convex front; flat back; foliate, annulets and pellets in the field; incomplete, a circular-section simple looped shank with a small section broken off, D 26.25mm; shank remnant L 7mm. Possibly a Mercers' Company livery button, perhaps depicting the Mercers' Maiden (an image that varied from the sublime to the ridiculous throughout time). In 1634 the Heralds of England acknowledged this image as the Mercers' Company Arms, sometimes known as the so-called 'Indian Queen' or 'Mary among clouds'. The Company has no record that this image explicitly represents the Virgin Mary; however, records concerning the dedication of the Drapers' Company would suggest strongly for such an association (pers. comm. Donna Marshall). Notwithstanding, this button is possibly a form of Billie and Charlie (pers. comm. Nick Griffiths). *Leicestershire.*

305. Discoidal; solid; convex front; flat back; undecorated; a lateral rim; a circular-section short stem and a circular-section loop; D 15mm; shank L c.8mm. *South-East Lincolnshire.*

306. Discoidal; solid; convex front; flat back; a hatched quatrefoil on a hatched field; a lateral rim; a circular-section simple looped shank; D 18.5mm; shank L 9mm. *Nottinghamshire.* UKDFD 5756.

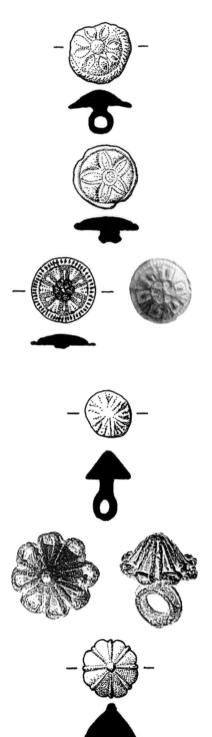

307. Discoidal; solid; convex front; flat back; an octofoil within a lateral cabled border; a circular-section short stem and a circular-section loop; incomplete, edge ragged; D 18mm; shank L 8mm. *South Somerset.*

308. Discoidal; solid; convex front; flat back; a cinquefoil within a circle; a lateral border; incomplete; edge slightly ragged and circular-section short stem, circular-section loop broken off; D 18mm; shank remnant L 3mm. *South-West Dorset.*

309. Discoidal; solid; convex front; flat back; a sexfoil with alternate radiating fan-shapes and globular-ended arms; a hatched lateral rim; a mould line on the back; incomplete, a circular-section simple looped shank broken off; D 17.5mm; shank remnant L 0.5mm. *Staffordshire.* UKDFD 9719.

310. Discoidal; solid; conical front; flat back; a multi-spoked wheel; a circular-section long stem and a circular-section loop; D 12mm; shank L 11mm. *South Somerset.*

311. Octofoil; solid; convex front; flat back; a central pellet, the ends of the recessed petals cant upwards; a mould line on the back and the shank; a circular-section short stem and a circular-section loop. *River Thames foreshore, London.*

312. Octofoil; solid; convex front; flat back; a central pellet and an octofoil, incomplete, shank broken off; D 15mm; shank remnant L 1.7mm. *East Devon.*

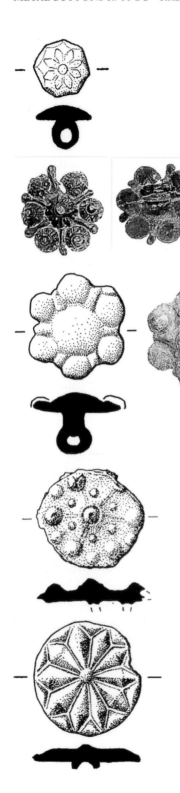

313. Octagonal; solid; convex front; flat back; an octofoil within a cusped octofoil, a mould line on the back; a circular-section simple looped shank; D 15.5mm; shank L 8mm. *East Yorkshire.*

314. Sexfoil; solid; convex front; flat back; each petal has a boss with a central pellet and a circle within voided lines; an offset circular-section simple looped shank; torn rim; (note anaerobic gilding). *River Thames foreshore, London.*

315. Sexfoil; solid; convex front; slightly convex back; a central boss bordered with six alternate smaller bosses and sepals; a circular-section long stem with a circular-section loop; D 27mm; shank L 11mm. *South Somerset.*

316. Discoidal; flat; a central pellet within a recessed circle; a recessed six-spoked wheel, each spoke terminating in a pellet, a large and small pellet in each angle; incomplete, edge ragged and an offset circular-section shank broken off; D 27mm. *East Devon.* After Read 1995, no. 372, erroneously described as a mount.

317. Discoidal; flat; a lateral edge; a multifoil with a central pellet; incomplete, a small section of edge ragged and a circular-section simple looped shank broken off; D 30mm; shank remnant L 1.8mm. *South Somerset.*

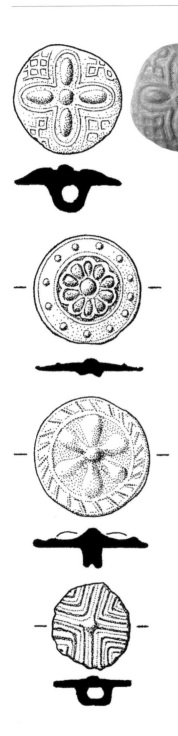

318. Discoidal; flat; a quatrefoil with a central pellet on a hatched field; a mould line on the back; a circular-section simple looped shank; D 29.5mm; shank L 7mm. *Berwickshire, Scotland.* UKDFD 10743.

319. Discoidal; flat; a central pellet and a multifoil within two concentric circles infilled with a band of small pellets; a mould line on the back; incomplete, a circular-section simple looped shank broken off; D 29mm; shank remnant L 1.8mm. *County Durham.* UKDFD 5577.

320. Discoidal; flat; a central pellet and a sexfoil within two concentric circles infilled with hatching; a mould line on the back; incomplete, a circular-section stem, circular-section loop broken off; D 31mm; stem remnant L 5.9mm. *Lincolnshire.* UKDFD 4258

321. Discoidal; flat; a saltire with a central pellet; three chevrons in each quadrant; a mould line on the back; a circular-section wide simple looped shank; D 21mm; shank L 5mm. *County Durham.* UKDFD 5397.

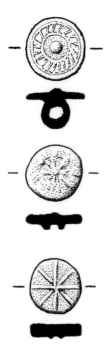

322. Discoidal; flat; a central pellet within two concentric circles infilled with hatching; a mould line on the back; a circular-section simple looped shank; D 16.5mm; shank L 7mm. *East Yorkshire.*

323. Discoidal; flat; a much abraded eight-spoked wheel with a central pellet, each spoke terminates in pellet; incomplete, a circular-section simple looped shank broken off; D 15.5mm; shank remnant L 1mm. *South Somerset.*

324. Discoidal; flat; an eight-spoked wheel; incomplete, a circular-section simple looped shank broken off; D 14.5; shank remnant L 0.5mm. *South-West Dorset.*

325. Discoidal; flat; four recessed hearts form a plain cross and a quatrefoil; four small pellets in the border; a small hole is probably a casting fault; a circular-section simple looped shank; D 26mm; shank L 8mm. *River Thames foreshore, London.*

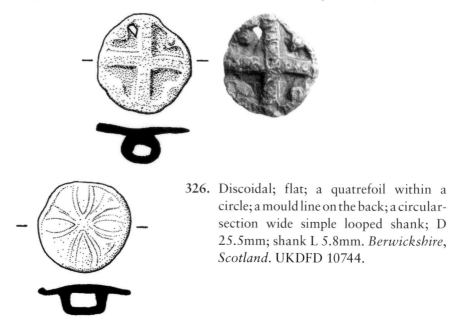

326. Discoidal; flat; a quatrefoil within a circle; a mould line on the back; a circular-section wide simple looped shank; D 25.5mm; shank L 5.8mm. *Berwickshire, Scotland.* UKDFD 10744.

84

327. Discoidal; flat; a male profile bust left; C between two pellets left; R between two pellets right; either King Charles I or King Charles II; a mould line on the back; incomplete, a circular-section stem, circular-section loop broken off; D 31mm; shank L 7.5mm. *West Suffolk*. After Morley 2004.

328. Discoidal; flat; a facing bust of a woman with (?) long-hair or (?) wearing a bonnet bordered with pellets; a mould line on the back; incomplete, a circular-section simple looped shank partially broken off; D 20mm; shank remnant L 8mm. *North Yorkshire*.

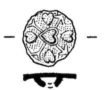

329. Discoidal; flat; a saltire formed from two arrows with a central heart, a heart in each quadrant; incomplete, edge ragged and a circular-section wide simple looped shank partially broken off; D 14.5mm; shank L 3mm. *Cambridgeshire*. UKDFD 4914.

330. Discoidal; flat; a sexfoil and a central boss, a pellet between each petal, within a circle; bordered with panels divided by wedges, alternate panels contain bosses or cross-hatching and three small pellets; a mould line on the back; a circular-section simple looped shank; rust staining on the front and back; D 25mm; shank L *c*.6mm. *West Suffolk*.

331. Discoidal; flat; a recessed circle within a beaded border; incomplete, a circular-section stem, loop broken off; a mould line on the back; D 9.5mm; shank remnant L *c*.2mm. *South-East Dorset*.

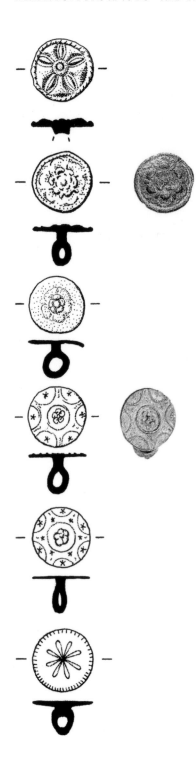

332. Discoidal; flat; a cinquefoil within a cabled border; incomplete, a circular-section stem, loop broken off; D 17mm; stem L 2.5mm. *East Devon.*

333. Discoidal; flat; a double six-petalled rose with sepals within a circle; a mould line on the back; a circular-section long stem and a circular-section loop; D 17mm; shank L 9mm. *South-West Wiltshire.*

334. Discoidal; flat; a double five-petalled rose; a circular-section long stem and a circular-section loop; D 15mm; shank L 9mm. *South Somerset.*

335. Discoidal; flat; a central cinquefoil within a circle bordered with semicircles containing five-pointed stars; a circular-section long stem and a circular-section loop; D 16mm; shank L 10mm. *South Somerset.*

336. Discoidal; flat; comparable with no. 335 but with an additional circle of five-pointed stars; a circular-section long stem and a circular-section loop; D 16mm; shank L 9mm. *South Somerset.*

337. Discoidal; flat; an octofoil bordered with radiating lines; a circular-section simple looped shank; D 17mm; shank L 7mm. *South Somerset.*

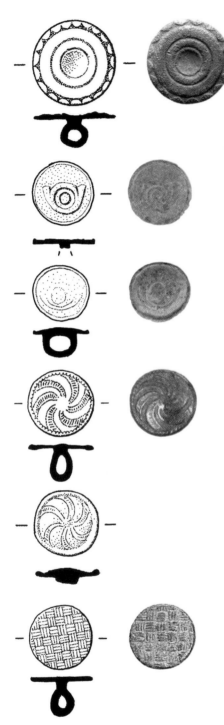

338. Discoidal; flat; a recessed centre; three concentric circles within an engrailed border; a mould line on the back; a circular-section short stem and a circular-section loop; D 23mm; shank L 7mm. *River Thames foreshore, London.*

339. Discoidal; flat; a raised rim; (?) two conjoined volutes or trumpets; incomplete, a circular-section short stem, loop broken off; D 18.5mm; stem remnant L 1.5mm. *South Somerset.*

340. Discoidal; flat; a raised rim; much abraded decoration perhaps comparable with no. 339; a circular-section very wide simple looped shank; D 15mm; shank L 6mm. *South Somerset.*

341. Discoidal; flat; a clockwise hatched six-bladed impeller within a beaded border; a mould line on the back; a circular-section short stem and a circular-section loop; D 17.5mm; shank L 9mm. *South Somerset.*

342. Discoidal; flat; a clockwise hatched six-bladed impeller within a recessed circle which creates a raised rim; a mould line on the back; incomplete, shank broken off; D 18mm; shank remnant L 2mm. *Lancashire.*

343. Discoidal; flat; basket weave; a circular-section stem and a circular-section loop; D 17mm; shank L 9mm. *South Somerset.*

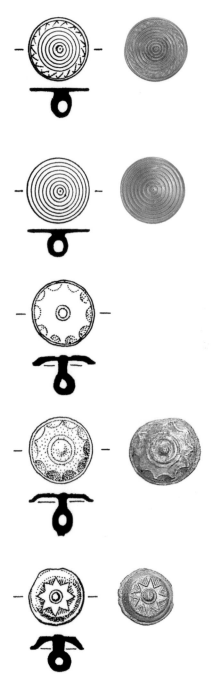

344. Discoidal; flat; a central pit and seven engraved – possibly engine-turned – concentric circles, a band of chevrons between the outside two; a circular-section very short stem and a circular-section loop; D 17mm; shank L 7mm. *South-West Wiltshire.*

345. Discoidal; flat; a central pit and eight engraved – possibly engine-turned – concentric circles; a circular-section very short stem and a circular-section loop; D 18mm; shank L 6mm. *South-West Wiltshire.*

346. Discoidal; flat; a central pellet within a circle; a downwards canted cusped border and lateral rim; a circular-section long stem and a circular-section loop; D 18mm; shank L 9mm. *South Somerset.*

347. Discoidal; flat; a central pellet within two concentric circles; a downwards canted and cusped border and lateral rim; a mould line on the back; a circular-section long stem and a circular-section loop with a casting sprue; D 18mm; shank L 10mm. *South-West Wiltshire.*

348. Discoidal; flat; a central pellet and circle within an eight-pointed star on a hatched field; a downwards canted and lateral rim; a mould line on the back; a circular-section long stem and a circular-section loop; D 15mm; shank L 8.5mm. *South Somerset.*

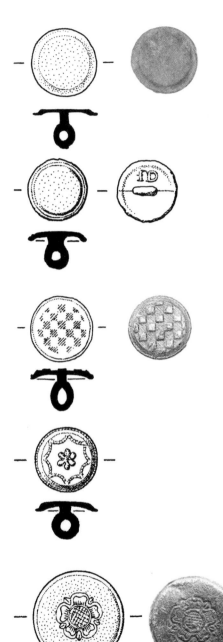

349. Discoidal; flat; undecorated; a downwards canted and lateral rim; a circular-section long stem and a circular-section loop with a casting sprue; D 18mm; shank L 10mm. *South Somerset.*

350. Discoidal; flat; undecorated; a raised border and downwards canted rim; an in-relief maker's mark ID and a mould line on the back; a circular-section short stem and a circular-section loop; D 17mm; shank L 9mm. *South-West Wiltshire.*

351. Discoidal; flat; four bands of contiguous lozenges on a hatched field; a downwards canted and lateral rim; a mould line on the back; a circular-section short stem and a circular-section loop; D 17mm; shank L 10mm. *North Dorset.*

352. Discoidal; flat; a sexfoil within two concentric circles, the inner cusped and the outer linear; a downwards canted and hatched lateral rim; a mould line on the back; a circular-section short stem and a circular-section loop; D 17.4mm; shank L 9.3mm. *South Somerset.*

353. Discoidal; flat; a five-petalled rose; a downwards canted and lateral rim; a mould line on the back; shank broken off; D 25.5mm. *River Thames foreshore, London.*

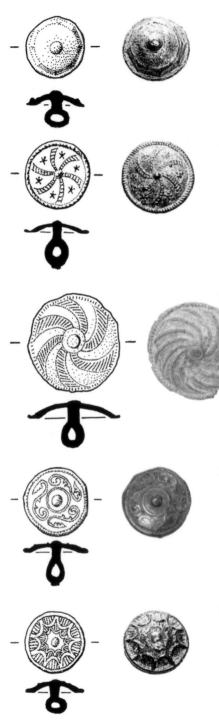

354. Discoidal; convex front; concave back; a central pellet within an octagon with raised edges; a lateral rim; a circular-section long stem and a circular-section loop; D 17mm; shank L 8mm. *South Somerset.*

355. Discoidal; convex front; concave back; a central pellet and an anticlockwise hatched six-bladed impeller, a five-pointed star in each segment, within a cabled lateral rim; a mould line on the back; a circular-section long stem and a circular-section loop with a casting sprue; D 18mm; 11.5mm. *South Somerset.*

356. Discoidal; convex front; concave back; a central pellet and a clockwise hatched six-bladed impeller, a hatched lateral rim; a circular-section long stem and a circular-section loop; D 26mm; shank L 11mm. *East Devon.*

357. Discoidal; convex front; concave back; a central pellet and circle surrounded by three S-scrolls within a cabled lateral rim; a mould line on the back; a circular-section long stem and a circular-section loop with a casting sprue; D 18mm; shank L 10mm. *South Somerset.*

358. Discoidal; convex front; concave back; a central pellet within two multi-pointed stars on a hatched field; a lateral rim; a circular-section long stem and a circular-section loop; D 16mm; shank L 8mm. *South Somerset.*

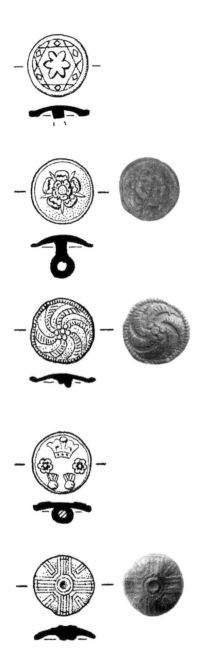

359. Discoidal; convex front; concave back; a sexfoil within a six-blunt-pointed star, a lozenge in each blunt point; a lateral rim; incomplete, a circular-section stem, loop broken off; D 18mm; shank remnant L 2mm. *South Somerset.*

360. Discoidal; convex front; concave back; a five-petalled rose with sepals; an abraded cabled border and a lateral rim; a circular-section long stem and circular-section loop; a mould line on the back; D 17mm; shank L 10mm. *Gloucestershire.* UKDFD 16989.

361. Discoidal; shallow convex front; shallow concave back; an anticlockwise hatched six-bladed impeller with a central sexfoil; a hatched upwards canted rim; incomplete, shank broken off; D 18mm; shank remnant L 2.3mm. *River Thames foreshore, London.*

362. Discoidal; shallow convex front; shallow concave back; two thistles surmounted by two six-petalled roses and a crown, within a circle; a circular-section simple loop retains rust; D 16mm; shank L 4mm. *County Durham.* UKDFD 4721.

363. Discoidal; convex front; concave back; a central pit within a raised circle and a six-stranded cross; three oblique lines in each quadrant; incomplete, a circular-section simple looped shank broken off; D17mm; shank remnant L 2mm. *River Thames foreshore, London.*

364. Discoidal; convex front; concave back; an M over (?) B on a six-pointed wavy star (possibly representing a royalist or Yorkist badge) on a hatched field; vestige of a mould line on the back; a circular-section simple looped shank; D 25mm; shank L uncertain. *East Yorkshire.* UKDFD 8224.

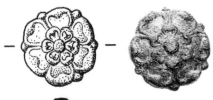

365. Sexfoil; convex front; concave back; a double six-petalled rose with sepals; a circular-section long stem and a circular-section loop; D 22mm; shank L 11mm. *South-West Wiltshire.*

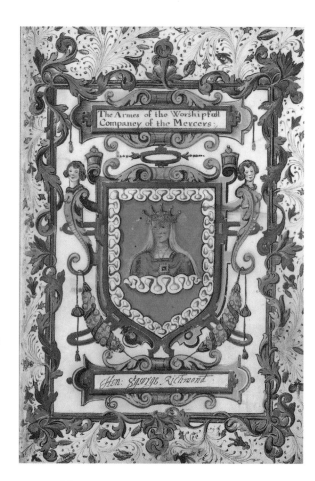

Fig 15. Mercers' Maiden from the Warden's Book of Arms 1634, reproduced by courtesy of the Mercers' Company.

Cast one-piece silver button with integral undrilled shank
Circa late 17th early 18th century.

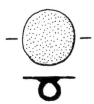

366. Discoidal; flat; undecorated; a circular-section simple looped shank; D 14.5mm; shank L 5mm. *South-West Wiltshire.*

Composite two-piece cast copper-alloy buttons with integral drilled shanks
All are *c.*17th century but continued throughout the 18th. Components are soldered together. Unless otherwise stated, decoration is engraved.

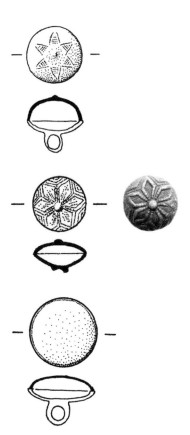

367. Biconvex; hollow; a central pellet within a six-pointed star; each point is hatched; a rectangular-section rounded shank integral with the back; D 15mm; shank L 4.5mm. *South Somerset.*

368. Biconvex; hollow; a central pellet within a sexfoil on a hatched field; incomplete, a rectangular-section shank integral with the back broken off; D 14.5mm; shank remnant L 1mm. *River Thames foreshore, London.*

369. Biconvex; hollow; undecorated; a white-metal coated front with a lateral rim; a rectangular-section rounded shank integral with the back; D 18mm; shank L 6mm. *South-West Wiltshire.*

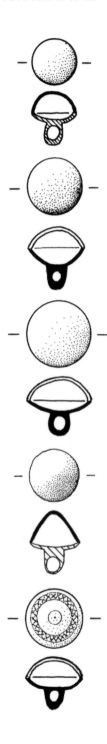

370. Biconvex; hollow; undecorated; a rectangular-section rounded shank integral with the back; D 12mm; shank L 5.5mm. *South-East Lincolnshire.*

371. Biconvex; hollow; undecorated; a rectangular-section rounded shank integral with the back; D 16mm; shank L 6mm. *South Somerset.*

372. Biconvex; hollow; undecorated; a rectangular-section rounded shank integral with the back; D 19mm; shank L 7mm. *South Somerset.*

373. Biconvex; hollow; undecorated; a rectangular-section rounded shank integral with the back; D 14mm; shank L 6mm. *South Somerset.*

374. Biconvex; hollow; a central circle-and-dot motif and two concentric circles infilled with cross-hatching; a rectangular-section rounded shank integral with the back; D 16mm; shank L 6mm. *South Somerset.*

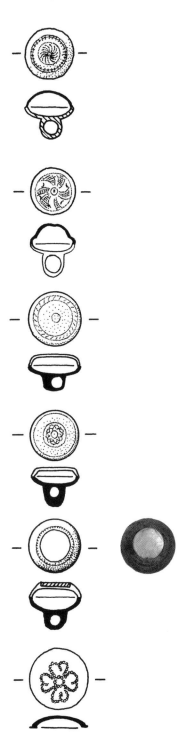

375. Biconvex; hollow; a central clockwise fifteen-bladed impeller within three concentric circles, the outer two infilled with hatching; a rectangular-section rounded shank integral with the back; D 13mm; shank L 5mm. *South-East Lincolnshire.*

376. Biconvex; hollow; a central circle-and-dot motif and a clockwise hatched six-bladed impeller within a circle and a lateral rim; a wedged-section rounded shank integral with the back; D 12mm; shank L 6mm. *East Devon.*

377. Sub-biconvex; flat top; hollow; three concentric circles, with hatching between the outer two; a rectangular-section trapezoid shank integral with the back; D 15mm; shank L 4.5mm. *South Somerset.*

378. Sub-biconvex; flat top; hollow; a central multifoil within two concentric circles; a rectangular-section rounded shank integral with the back; D 14mm; shank L 6mm. *South Somerset.*

379. Sub-biconvex; hollow; an inset mother-of-pearl roundel bordered with tiny contiguous triangles; a rectangular-section rounded shank integral with the back; D 15mm; shank L 4.5mm. *South Somerset.*

380. Discoidal; convex; a quatrefoil; a lateral rim; incomplete, front of a biconvex hollow button, probable integral shank type; D 16mm. *East Devon.*

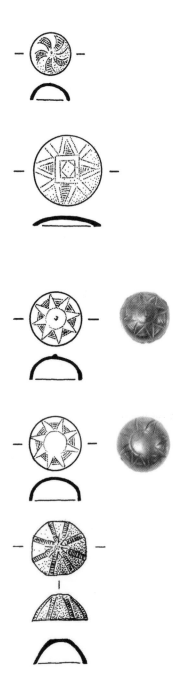

381. Discoidal; convex; a central pit and a clockwise hatched five-bladed impeller, incomplete, front of a biconvex hollow button, probable integral shank type; D 12mm. *South Somerset.*

382. Discoidal; shallow convex; a central lozenge within two concentric squares and a voided hatched four-pointed star, two oblique lines in each quadrant; resembles a quatrefoil with sepals; incomplete, front of a biconvex hollow button, probable integral shank type; D 19mm. *East Devon.*

383. Discoidal; convex; a central pellet within a circle and an eight-pointed star, hatched in the angles; incomplete, front of a biconvex hollow button, probable integral shank type; D 14mm. *South Somerset.*

384. Discoidal; convex; a central circle and an eight-pointed star, hatched in the angles; incomplete, front of a biconvex hollow button, probable integral shank type; D 14.5mm. *East Devon.*

385. Discoidal; convex; a voided eight-spoked wheel, each spoke infilled with hatching; random oblique lines in the angle points; incomplete, front of a biconvex hollow button, probable integral shank type; D 15mm. *South-West Dorset.*

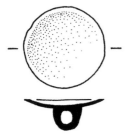

386. Discoidal; concave; a rectangular-section trapezoid shank with a rounded end; incomplete, back of a biconvex hollow button; D 22mm; shank L 6mm. *South-East Dorset.*

Composite two-piece cast copper-alloy button or cloak-fastener with integral drilled shank

Whether a true button or a cloak-fastener is uncertain. A 17th-century context of the River Thames foreshore, London revealed a heavily white-metal coated parallel, and another example is in the Ashmolean Museum, Oxford (several others from inland have come to light since publication of edition one). Decoration is moulded-in-relief and openwork.

387. Discoidal; hollow; convex front; flat back; an octofoil with S-spirals formed from asymmetrical openwork in each petal; a cusped rim; the back has openwork forming an octofoil within a circle of asymmetrical openwork; an integral central vertical pin rises from the inside of the back which is then riveted via a hole in the front; sporadic white-metal coating overall; a rectangular-section trapezoid shank with a rounded end; D 29mm; shank L 6mm. *South Somerset.*

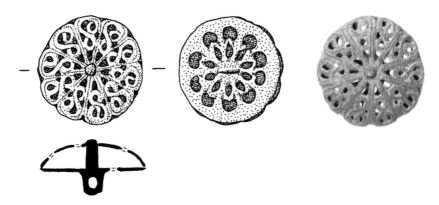

Composite two-piece cast copper-alloy and sheet iron button with integral drilled shank
Circa late 17th – 18th century. Whether a true button or a cloak-fastener is uncertain. Decoration is moulded-in-relief.

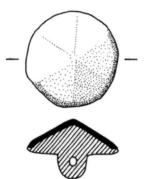

388. Discoidal; solid; convex front; flat back; six triangular facets, front white-metal coated; corroded iron back with a rectangular-section rounded shank; found with a remnant of a separate drawn iron wire link retained in the hole, now lost; D 26mm; shank L 5mm. *East Devon*.

Composite two-piece cast copper-alloy and lead/tin alloy button with integral undrilled shank
Circa 17th century. Whether a true button or a cloak-fastener is uncertain. Decoration is moulded-in-relief.

389. Discoidal; shallow convex front; shallow concave back; the lead/tin alloy front, which is somehow affixed to the copper-alloy back, has four recessed heart-shapes which form a saltire and a quatrefoil within a circle; incomplete, edge ragged, a circular-section long stem, circular-section loop broken off; D 29mm; shank remnant L 8mm. *South-East Lincolnshire*.

Composite two-piece cast copper-alloy buttons with separate embedded drawn copper-alloy wire shanks
Nos 398, 411 are perhaps 16th century, the remainder are *c*.17th century. Nos 405–10 are from a 17th-century context. These buttons may have a high-tin content or a polished tin coating, both of which may create an appearance of silver. Silver examples possibly exist. Unless otherwise stated, decoration is moulded-in-relief.

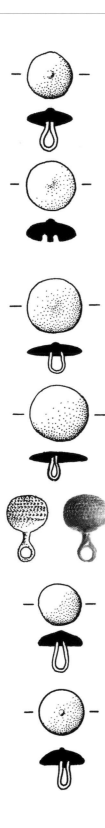

390. Biconvex; solid; undecorated apart from a central pellet; a circular-section simple looped shank; D 12mm; shank L 8mm. *South-West Wiltshire.*

391. Biconvex; solid; undecorated apart from a central pellet; incomplete, the absence of rust staining around the circular shank-holes implies a probable circular-section copper-alloy wire simple looped shank broken off; D 14mm. *South Somerset.*

392. Biconvex; solid; undecorated apart from a central pellet; a circular-section simple looped shank; D 15mm; shank L 5mm. *South-West Dorset.*

393. Biconvex; solid; undecorated; a circular-section simple looped shank; D 16mm; shank L 4.5mm. *South-West Wiltshire.*

394. Biconvex; solid; multiple concentric dotted circles; a circular-section long stem and a circular-section loop; gilded overall; D 11mm; shank L 8mm. *South-West Dorset.*

395. Discoidal; solid; convex front; flat back; undecorated apart from a shallow central pellet; a circular-section simple looped shank; D 12mm; shank L 8mm. *South-West Wiltshire.*

396. Discoidal; solid; convex front; flat back; undecorated apart from a central pellet; a circular-section simple looped shank; D 12mm; shank L 7mm. *South-West Wiltshire.*

99

397. Discoidal; solid; convex front; flat back; high-tin content; undecorated apart from a central pellet; a circular-section simple looped shank; D 13.5mm; shank L 7mm. *South-East Dorset.*

398. Discoidal; solid; convex front; flat back; a central pellet within five concentric circles of small pellets; a circular-section simple looped shank; D 10mm; shank L 10mm. *Cambridgeshire.*

399. Discoidal; solid; convex front; flat back; a central pellet and an eight-spoked wheel, a pellet between each spoke; bordered with alternate large and small pellets; a downwards canted lateral rim; white-metal coated overall; incomplete, edge ragged and the absence of rust staining around the circular shank-holes implies a probable circular-section copper-alloy wire simple looped shank broken off. D 25mm. *South Somerset.*

400. Discoidal; solid; convex front; flat back; undecorated; a circular-section simple looped shank; D 14mm; shank L 7mm. *South Somerset.*

401. Discoidal; solid; convex front; flat back; undecorated; a circular-section simple looped shank; D 12.5mm; shank L 4.5mm. *South Somerset.*

402. Discoidal; solid; convex front; flat back; a polished white-metal coating; undecorated; a circular-section simple looped shank; D 17mm; shank L 6mm. *Cambridgeshire.*

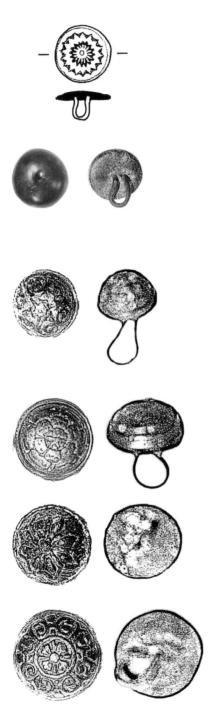

403. Discoidal; solid; shallow convex front; flat back; an engraved central annulet within two concentric multifoils and a circle; a circular-section simple looped shank; D 15mm; shank L 5.5mm. *East Devon.*

404. Discoidal; solid; convex front; flat back; undecorated apart from a central pit; sporadic white-metal coating on the back; a circular-section simple looped shank; D 12mm; shank L 6.5mm. *South-West Wiltshire.*

405. Discoidal; solid; convex front; flat back; an elaborate quatrefoil and central pellet with a stylised fleur-de-lis in each angle; a circular-section simple looped shank; (note anaerobic gilding). *River Thames foreshore, London.*

406. Discoidal; solid; convex front; flat back; a double six-petalled rose within a circle; a circular-section simple looped shank; (note anaerobic gilding). *River Thames foreshore, London.*

407. Discoidal; solid; convex front; flat back; an octofoil and a central pellet within a cabled border; incomplete, shank broken off; (note anaerobic gilding). *River Thames foreshore, London.*

408. Discoidal; solid; convex front; flat back; a central four-petalled rose with sepals within a circle bordered with spirals and pellets; a circular-section simple looped shank; (note anaerobic gilding). *River Thames foreshore, London.*

101

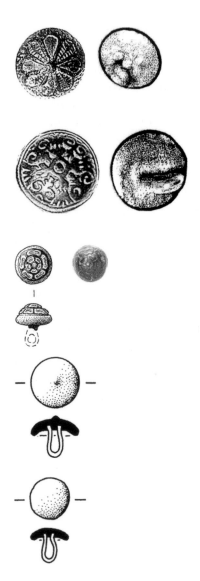

409. Discoidal; solid; convex front; flat back; an octofoil and a central pellet within a cabled border; a circular-section simple looped shank; (note anaerobic gilding). *River Thames foreshore, London.*

410. Discoidal; solid; convex front; flat back; an elaborate cross with an elaborate trefoil in each quadrant; a circular-section simple looped shank; (note anaerobic gilding). *River Thames foreshore, London.*

411. Discoidal; solid; convex front; flat back; a double five-petalled rose; sporadic black paint on the front; D 10mm; shank remnant L 1mm. *North Dorset.*

412. Discoidal; convex front; concave back; undecorated apart from a central pellet; sporadic white-metal coating overall; a circular-section simple looped shank; D 13.5mm; shank L 8mm. *South-East Dorset.*

413. Discoidal; convex front; concave back; undecorated; white-metal coated overall; a circular-section simple looped shank; D 11mm; shank L 8mm. *South-East Dorset.*

Composite two-piece cast copper-alloy buttons with separate embedded drawn iron wire shanks

All are *c*.17th century. Nos 418-19 are from a *c*.17th-century context. Unless otherwise stated, decoration is engraved and/or punched.

414. Discoidal; flat; a voided saltire formed from addorsed chevrons, annulets within radiating lines in the centre and in each quadrant, bordered with dots; incomplete, a circular-section simple looped shank broken off; D 18mm; shank remnant L 4mm. *South Somerset.*

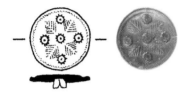

415. Discoidal; flat; a cross formed from five circle-and-dot motifs with engrailed edges, a veined leaf in each quadrant, resembles a quatrefoil within a circle of zigzags; a circular-section simple looped shank broken off; D 17mm; shank remnant L 2mm. *Lincolnshire.* UKDFD 3595.

416. Discoidal; flat; a central annulet and a clockwise multi-dotted four-bladed impeller and crescents within a circle; a probable circular-section simple looped shank broken off; D 15.5mm; shank remnant L 3mm. *East Devon.*

417. Discoidal; solid; convex front; flat back; a central circle-and-dot motif and six groups of dots inside a six-pointed star, within a circle; incomplete, a circular-section short stem and a circular-section loop broken off; D 15mm; shank remnant L 5mm. *South-West Wiltshire.*

418. Discoidal; solid; convex front; flat back; a recessed octofoil with a central pellet, sporadic gilding or gold-leaf on several petals; a mould line on the back; a circular-section stem and circular-section loop retaining two fragments of a broken eye-section of a Read Class A, Type 1 white-metal coated drawn copper-alloy wire clasp. *River Thames foreshore, London.*

103

419. Discoidal; solid; convex front; flat back; moulded-in-relief Prince of Wales' feathers and the motto ICH DIEN [I serve] bordered with a wreath; a mould line on the back; incomplete, a probable circular-section simple looped shank broken off; (note anaerobic gilding). *River Thames foreshore, London.*

Composite two-piece cast lead/tin alloy buttons with separate embedded drawn copper-alloy wire shanks

Nos 423, 431-33, 436-39 are from a *c*.17th-century context. No. 426 is perhaps early 18th century, and no. 435 late 17th early 18th. No. 439 is white-metal coated or is possibly silver. Unless otherwise stated, decoration is moulded-in-relief.

420. Discoidal; flat; a central cinquefoil within two concentric circles and a six-armed blunt-pointed star; bordered with small cinquefoils and crescents; incomplete, a probable circular-section simple looped shank broken off; D 17mm; shank remnant L 1mm. *East Devon.*

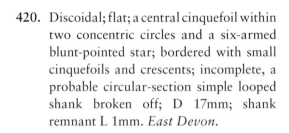

421. Discoidal; flat; solid; undecorated; the edge tapers with a lateral rim at the shank end; a mould line on the back; incomplete, the absence of rust staining around the circular shank-holes implies the missing shank was circular-section copper alloy wire; D 15.5mm. *Gloucestershire.*

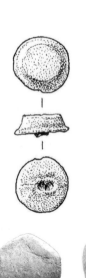

422. Discoidal; flat; wedged-section; undecorated; a gouge to one side of the shank is perhaps due to a wire clip; a circular-section simple looped shank; D 20mm; shank L 4mm. *Wiltshire.* WHM, acc. no. 1995.125.1.

423. Discoidal (virtually plano-convex); solid; flat front; convex back; a central pellet within a sexfoil bordered with pellets; incomplete, a circular-section simple looped shank broken off; D 12.5mm; shank remnant L 1.5mm. *River Thames foreshore, London.*

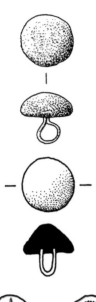

424. Discoidal; solid; convex front; flat back; undecorated; a circular-section simple looped shank; D 14.5mm; shank L 7mm. *South-West Wiltshire.*

425. Discoidal; solid; convex front; flat back; undecorated; a circular-section simple looped shank; D 14mm; shank L 6mm. *South-East Dorset.*

426. Discoidal; solid; convex front; flat back; a sexfoil with sepals on a hatched field; on the back are two concentric beaded circles with remnants of an inscription C[...]IOH; incomplete, edge ragged, the absence of rust staining around the circular shank-holes implies the missing shank was circular-section copper alloy wire; D 16mm. *South-East Dorset.*

427. Discoidal; solid; convex front; flat back; a blunt-pointed voided triangle with hatching inside at each angle; incomplete, a circular-section simple looped shank broken off; D 12.6mm; shank remnant L 1mm. *East Devon.*

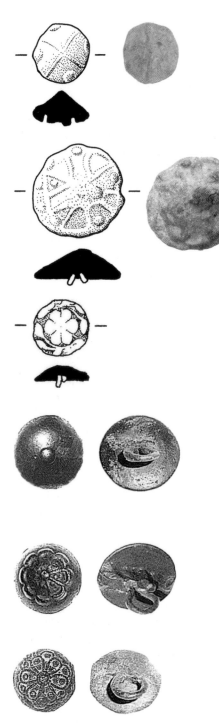

428. Discoidal; solid; convex front; flat back; a saltire with a pellet in each quadrant; the absence of rust staining around the circular shank-holes implies the missing shank was circular-section copper alloy wire; D 16mm. *South Somerset.*

429. Discoidal; solid; convex front; flat back; pellets and contiguous triangles; incomplete, a circular-section simple looped shank, which is also exposed on the front, broken off; D 24.5mm; shank remnant L 2mm. *South Somerset.*

430. Discoidal; solid; convex front; flat back; a sexfoil within a wreathed border; incomplete, a circular-section simple looped shank broken off; D 16mm; shank remnant L 3mm. *North-West Dorset.*

431. Discoidal; solid; convex front; flat back; undecorated apart from a central pellet; a mould line on the back; a circular-section simple looped shank; (note anaerobic gilding). *River Thames foreshore, London.*

432. Discoidal; solid; convex front; flat back; a central pellet and an octofoil; a mould line on the back; a circular-section simple looped shank; (note anaerobic gilding). *River Thames foreshore, London.*

433. Discoidal; solid; convex front; flat back; a multifoil with two concentric central circles, a pellet within each petal; a circular-section simple looped shank; (note anaerobic gilding). *River Thames foreshore, London.*

434. Discoidal; shallow concave front; shallow convex back; four recessed linked roundels, each within a circle; a chevron between each roundel on a hatched field; resembles a quatrefoil with sepals; a mould line on the back; a circular-section simple looped shank; D 16mm; shank L 6mm. *Wiltshire.* UKDFD 20743.

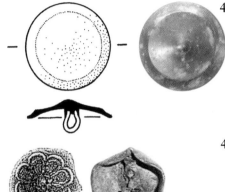

435. Discoidal; convex front; concave back; undecorated apart from a shallow central pellet within a concave downwards canted rim; a circular-section simple looped shank; D 24mm, shank L 6mm. *South Somerset.*

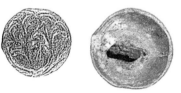

436. Discoidal; convex front; concave back; an octofoil with a central boss; a mould line on the back; incomplete, a probable circular-section simple looped shank broken off, misshapen; (note anaerobic gilding). *River Thames foreshore, London.*

437. Discoidal; convex front; concave back; foliate; a circular-section simple looped shank; (note anaerobic gilding). *River Thames foreshore, London.*

438. Discoidal; convex front; concave back; a central solid square within a square and a multifoil; a mould line on the back; incomplete, shank broken off; (note anaerobic gilding). *River Thames foreshore, London.*

439. Discoidal; convex front; concave back; undecorated; a circular-section simple looped shank; (note anaerobic gilding). *River Thames foreshore, London.*

107

Composite two-piece cast lead/tin alloy buttons with separate embedded drawn iron wire shanks

Nos 440-48 are *c.*17th century. No. 449 is *c.* mid-17th – 18th century. The crudeness of nos 441, 443-44 suggests they are perhaps home-made. Unless otherwise stated, decoration is moulded-in-relief.

440. Discoidal; flat; a pellet and circle within an eight-pointed star, hatched in the angles; a chamfered edge; incomplete, edge ragged and flaking, a probable circular-section simple looped shank broken off; D *c.*20mm; shank remnant L *c.*4mm. *North Dorset.*

441. Discoidal; flat; a saltire, each arm reaches the edge, with a central pellet; each quadrant is recessed thus forming a raised border with pellets; incomplete, a probable circular-section simple looped shank broken off, which is also exposed on the front; maximum D *c.*38mm; shank remnant L *c.*1.5mm. *North Dorset.*

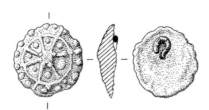

442. Discoidal; solid; convex front; flat back; a seven-spoked wheel with a pellet in each angle bordered with pellets, several spokes project and form segments; an irregularly cusped edge; an offset circular-section simple looped shank; D 34mm; shank L 8mm. *South-West Dorset.*

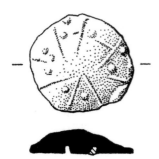

443. Discoidal; solid; convex front; flat back; a crude eight-spoked wheel and random pellets; incomplete, a probable circular-section simple looped shank broken off; D 29mm. *South-East Dorset.*

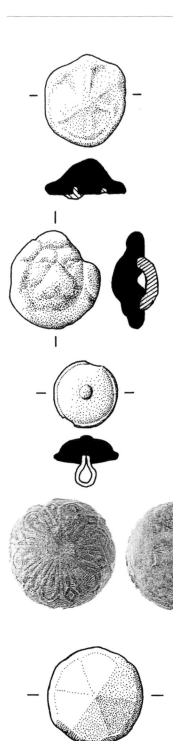

444. Discoidal; solid; convex front; flat back; a six-spoked wheel within a circle; incomplete, a probable circular-section simple looped shank broken off; D 24mm. *South-East Lincolnshire.*

445. Discoidal; solid; convex front; flat back with a central recess; an abraded facing mask of a lion bordered with segments; incomplete, edge slightly ragged; a circular-section simple looped shank; D 26mm; shank L 6mm. *East Devon.*

446. Sub-biconvex; solid; undecorated apart from a central pellet within a lateral rim; incomplete, a slightly ragged edge; a circular-section simple looped shank; D 18mm; shank L 6mm. *South-East Dorset.*

447. Biconvex; solid; a sexfoil with lozenge-headed sepals; a maker's mark (?) I C and a circle with an oblique line and a mould line on the back; incomplete, a probable circular-section simple looped shank broken off; D 30mm. *Wiltshire.* WHM, acc. no. 3.1980. After Giltsoff and Robinson. 1980.

448. Discoidal; convex front; concave back; high-tin content; six triangular facets with a faceted border; incomplete, a probable circular-section simple looped shank broken off; D 26mm. *East Devon.*

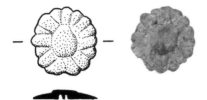

449. Multifoil; solid; convex front, flat back; incomplete, rust staining around the circular shank-holes implies the missing shank was circular-section iron wire; D 18mm. *South Somerset.*

Composite two-piece cast copper-alloy buttons or cloak-fasteners with separate soldered drawn iron wire shanks

All are *c.*17th century. Whether true buttons, cloak-fasteners or even mounts is uncertain. Unless otherwise stated, decoration is engraved and/or punched.

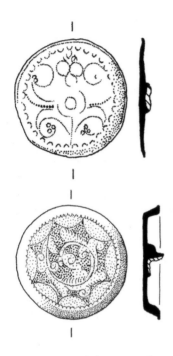

450. Discoidal; flat; curlicues and annulets bordered with crescents; a shallow lateral rim; incomplete, a probable circular-section simple looped shank broken off; D 32mm. *East Devon.*

451. Discoidal; flat front; concave back; curlicues and dots and an eight-pointed star within an engrailed circle; a downwards canted edge with a lateral rim; a white-metal coated front; incomplete, a probable circular-section simple looped shank broken off; D 31mm. *South-West Wiltshire.*

452. Discoidal; convex front; concave back; a moulded-in-relief central pellet within three concentric circles; a slightly upwards canted lateral rim; incomplete, a probable circular-section simple looped shank broken off; D 26.5mm. *South Somerset.*

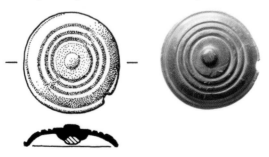

Composite two- or three-piece cast lead/tin alloy buttons with drawn iron wire or indeterminate shanks
All are *c*. late 16th – 17th century. No. 455 is from a *c*.16th – *c*.17th century context. Decoration is moulded-in-relief.

453. Biconvex; hollow; a central pellet and a circle within a sexfoil; two blow-holes in the back which is possibly soldered to the front; incomplete, two circular holes (one torn) in the back imply a missing circular-section simple looped shank which may have been soldered inside or outside; D 15mm. *Wiltshire*. WHM, acc. no. unknown.

454. Discoidal; hollow; convex front; flat back; a central pellet and a four-pointed star and circle within a multi-pointed star; incomplete, a separate back, which is probably soldered to the front, is squashed, and two ragged-edged holes imply either an integral shank broken-off or a separate wire simple looped shank torn out; D 27.5mm. *South-Somerset*.

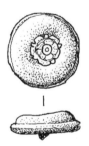

455. Discoidal; hollow; convex front; flat back; a double five-petalled rose with sepals; a lateral rim; incomplete; front slightly squashed, a circular-section simple looped shank broken off; D 20mm; shank remnant L 2mm. *River Thames foreshore, London*. UKDFD 22184.

Composite three-piece die-stamped sheet copper-alloy buttons with separate soldered drawn copper-alloy wire shanks
All are *c*.17th century, but probably extended into the 18th. Nos 459, 469-75 are from a *c*.17th-century context. Components are soldered together. Unless otherwise stated, decoration is engraved.

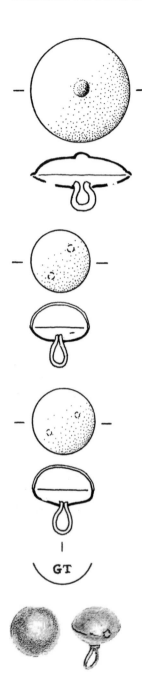

456. Biconvex; hollow; undecorated apart from a central die-stamped pellet and a lateral rim; two blow-holes in the back; a circular-section simple looped shank; D 27mm; shank L 6mm. *East Devon.*

457. Biconvex; hollow; undecorated; two blow-holes in the back; a circular-section simple looped shank; D 17mm; shank L 9mm. *South Somerset.*

458. Biconvex; hollow; undecorated; maker's mark GT stamped on the back; two blow-holes in the back; a circular-section simple looped shank; D 18mm; shank L 9mm. *South Somerset.*

459. Biconvex; hollow; undecorated; two blow-holes in the back; a short stem and a circular-section loop; (note anaerobic gilding). *River Thames foreshore, London.*

460. Biconvex; hollow; a die-stamped double five-petalled rose with sepals and a lateral rim; two blow-holes in the back; a hemispherical-section simple looped shank; D 27mm; shank L 7mm. *South-East Dorset.*

461. Biconvex; hollow; two concentric circles within a cinquefoil with sepals; two blow-holes in the back; a circular-section simple looped shank; D 29mm; shank L 5mm. *Wiltshire.* WHM, acc. no. 1996.13.

462. Biconvex; hollow; a quatrefoil with sepals within a circle; a circular-section simple looped shank; D 16mm; shank L 3mm. *South-West Wiltshire.*

463. Biconvex; hollow; a central saltire and wavy circle within two concentric circles, a nine-pointed star within a further two concentric circles; two blow-holes in the back; a circular-section simple looped shank; D 28mm; shank L 9mm. *South Somerset.*

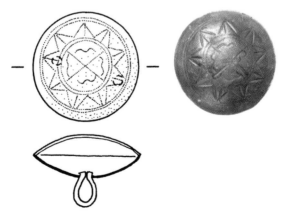

464. Biconvex; hollow; five concentric circles; two blow-holes in the back; a circular-section simple looped shank; D 28mm; shank L 10mm. *South Somerset.*

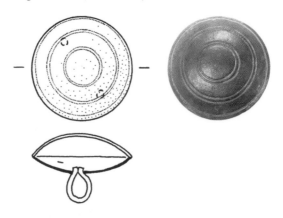

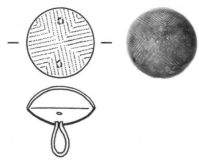

465. Biconvex; hollow; a cross formed from multi-chevrons; sporadic gilding on the front; two blow-holes in the back; a circular-section simple looped shank; D 19mm; shank L 9mm. *South Somerset.*

466. Biconvex; hollow; a clockwise hatched five-bladed impeller within a circle; two blow-holes in the back; a circular-section simple looped shank; D 16mm; shank L 8mm. *South Somerset.*

467. Biconvex; hollow; a sprig with three flowers; a circular-section simple looped shank; D 12mm; shank L 6mm. *South-West Wiltshire.*

468. Biconvex; hollow; faceted front; annulets and dots; gilded overall; a circular-section simple looped shank; D 12mm; shank L 4.5mm. *East Devon.*

469. Biconvex; hollow; an abraded sexfoil within two concentric circles; two blow-holes in the back; a circular-section short stem and a circular-section loop; (note anaerobic gilding). *River Thames foreshore, London.*

470. Biconvex; hollow; a central sexfoil within overlapping voided triangles forming a six-pointed star, two blow-holes in the back; a circular-section short stem and a circular-section loop; (note anaerobic gilding). *River Thames foreshore, London.*

471. Biconvex; hollow; a much abraded central pellet within a field of (?) small flowers; two blow-holes in the back; a circular-section simple looped shank; (note anaerobic gilding). *River Thames foreshore, London.*

472. Biconvex; hollow; a central much abraded sexfoil on a hatched field; two blow-holes in the back; a circular-section simple looped shank (note anaerobic gilding). *River Thames foreshore, London.*

473. Biconvex; hollow; a much abraded (?) flower with sepals on a hatched field; two blow-holes in the back; a circular-section simple looped shank; (note anaerobic gilding). *River Thames foreshore, London.*

115

474. Discoidal; hollow; convex front; flat back; undecorated; a white-metal coated front; two blow-holes in the back; a circular-section simple looped shank; (note anaerobic gilding). *River Thames foreshore, London.*

475. Discoidal; hollow; convex front; flat back; an abraded voided six-spoked wheel on a hatched field; two blow-holes in the back; a circular-section simple looped shank; (note anaerobic gilding). *River Thames foreshore, London.*

Composite four-piece repoussé sheet silver button with separate soldered drawn silver wire shank

Possibly not a button; however, stylistically its construction is perhaps similar to several gold buttons recovered from the Spanish vessel the *Girona*. Components are soldered together. The style of decoration on this object remained popular from the late 16th century through to the 19th, therefore its attribution is uncertain.

476. Biconvex; hollow; a double quatrefoil and sexfoil within a foliate border; the body comprises two hemispherical sections; incomplete, a missing separate sheet back-plate (probably with a blow-hole) and a separate circular-section simple looped shank broken off; D 12.5mm. *East Devon.*

Composite three-piece die-stamped sheet silver buttons with separate soldered drawn silver wire shanks

Nos 477-78 are c.17th century and no. 479 c.16th century. Components are soldered together. Unless otherwise stated, decoration is engraved.

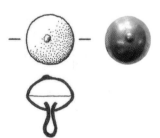

477. Biconvex; hollow; undecorated apart from a central die-stamped pellet; a circular-section simple looped shank; D 13mm; shank L 7mm. *East Devon.*

478. Biconvex; hollow; a sexfoil on a hatched field within two concentric circles; alternate hatched and plain panels between the circles; a hatched border; the back has two blow-holes and a maker's mark RS; a circular-section simple looped shank; D 12.5mm; shank L 4mm. *South-West Dorset*

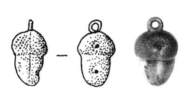

479. Figurative; a hollow acorn; two blow-holes in one side; a circular-section simple looped shank; D 18.5mm; L 12.4mm; shank L 2mm. *Hertfordshire.*

Composite two-piece cast silver button with separate embedded drawn silver wire shank
Possibly 17th century or early 18th. Decoration is moulded-in-relief.

480. Discoidal; flat; a central human facing mask within a twelve-rayed star on a hatched field; an in-relief maker's mark I L and a mould line on the back; incomplete, the absence of rust staining around two circular holes in the back implies the shank was probably circular-section silver wire; D 17mm. *River Thames foreshore, London.* UKDFD 19855.

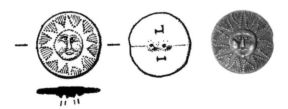

Composite silver openwork filigree buttons with separate soldered drawn silver wire shanks
Both are probably late 16th century. Components are soldered together. The primary manufactory element typical of these hollow buttons is drawn wire set in individual wire frames and is called openwork filigree, which is also known in gold or base-metal. When this wirework is soldered to a separate solid metal or woven wire mesh base, it is termed as ground-supported filigree. Other forms of filigree are a combination of the aforementioned two, and filigree to which another material, for example enamel or niello, is used to fill in the spaces.

Wires used to make filigree may be flat, thin strips, twisted or plain; square-section, twisted or plain; plaits or braids or stamped or incised beading. In traditional filigree, wires are always soldered together, never overlapped. In conjunction with some filigree, one or more gold, silver or base-metal granules, which often imitate pearls, provide additional decoration. Granules are affixed by the fusion process, a technique dating back to the Etruscans, although inferior modern work may be soldered.

Openwork filigree and granular decoration, apart from the air spaces, appears much the same as ground-supported filigree and granules found on silver, silver-gilt or base-metal beads, hooked-clasps, hollow biconvex-headed dress-pins, finger-rings, hanging-buttons and post-buttons (the latter are commonly known as collar studs) attributed to the late Middle Ages, the 16th and even 19th century.

Until recently, hollow biconvex-headed dress-pins of this type were erroneously classified as *c*.9th – *c*.10th-century Anglo-Saxon, and indeed some museums still exhibit them as such. Nonetheless, both the Anglo-Saxons and the Anglo-Scandinavians certainly did use ground-supported filigree on some of their jewellery.

In the Middle Ages filigree was also known as 'Venetian fashion', 'Venice work' or 'Damascus fashion': Venice was a centre for goldsmithing from at least the 14th century, especially for precious-metal openwork and ground-supported filigree buttons.

Prior to 1658 Skåne (Scania in English), Sweden, was part of Denmark, and like Scandinavia generally, has a long tradition of crafting base-metal, silver (both normally gilded), and probably gold, buttons, hanging-buttons, post-buttons, finger-rings, beads (especially paternoster beads) and hollow biconvex-headed dress-pins (and other items of jewellery) decorated with ground-supported filigree and granules.

This style of jewellery was known – perhaps to a lesser extant – in England from at least the early 15th century, but possibly became more fashionable in Elizabethan times, between 1559 and the 1580s and perhaps later, when the English and Swedish courts had close connections. On 1st October, 1559 Duke John of Finland arrived in England intent on persuading Queen Elizabeth I to marry his brother Prince Eric (who since 1556 had desired to marry Elizabeth), the King of Sweden's eldest son. Duke John stayed at the English court until April 1560, when he left for home, unsuccessful in pressing the suit of his brother. Reputedly, the Duke habitually scattered coinage – much of which turned out to be counterfeit – to the masses as he passed by. In 1565 Princess Cecilia of Sweden visited England to negotiate the marriage of her brother Eric, now king of Sweden, and Queen Elizabeth I. Among the Princess' entourage was 16-year-old Helena Snachenberg (one reference says she was Danish, which is correct if she came from Skåne), who soon afterwards married the Marquis of Northampton who died five years later. In 1580 Helena remarried, this time to Sir Thomas Gorges – a high-ranking member at the court of Queen Elizabeth I – of Longford Castle near Salisbury, Wiltshire. Helena became a favourite of Queen Elizabeth I and became her maid of honour. Is it too fanciful to believe that

Duke John, Princess Cecilia or Helena – or perhaps all three – brought with them Skåne-type jewellery as gifts to other courtiers, thereby restarting the fashion?

Whether openwork filigree or ground-supported filigree buttons, hanging-buttons, post-buttons, finger-rings, beads and hollow biconvex-headed dress-pins found in Britain are Scandinavian- or English-made is uncertain. Likewise, uncertainty surrounds whether they were produced at a single workshop – either Scandinavian or English – or widely dispersed English workshops, using the same patterns. Interestingly, the now ubiquitous in Britain hooked-clasps and hat-hooks having ground-supported filigree and granular decoration appear not to be Swedish (or other Scandinavian) in origin and several hooked-clasps stamped with a maker's mark now in the known record indicate they are almost certainly English, as are probably all others (see Read 2008).

Probable Elizabethan era, and later, silver, silver-gilt or base-metal hollow or solid biconvex-headed dress-pins, hanging-buttons, beads (and hooked-clasps) with moulded-in-relief imitative ground-supported filigree and granular decoration are also recorded. These are invariably of inferior workmanship and probably not Scandinavian.

481. Octofoil; biconvex; hollow; both halves have petals of drawn single strand plain wire, each of which encloses an S-spiral of twisted single strand drawn wire; a central imitative pearl granule; possible sporadic gilding overall; a hemispherical-section simple looped shank is soldered to the back; D 15.5mm; shank L 7mm. *South Somerset.*

482. Octofoil; biconvex; hollow; both halves have petals of drawn single strand plain wire, each of which encloses an S-spiral of twisted single strand drawn wire; both the front and back has a central hole through which passes the hemispherical-section stem, and a circular-section loop; the front end of the stem is riveted and forms a central imitative pearl; slightly squashed, abraded front; D 14mm; shank L 5mm. *South-West Devon.*

119

Composite three-piece silver infilled openwork filigree and sheet copper-alloy button with separate soldered drawn copper-alloy wire shank

Probably late 16th century. Components are soldered together. This button is unusual in that it is part copper alloy and part silver.

483. Multifoil; biconvex; hollow; the squashed and damaged silver front is drawn single plain strand wire petals, each of which contains an S-spiral of twisted single strand silver drawn wire; each loop of the S-spirals holds an imitative pearl granule and a similar but larger granule lies within a central annulet of twisted single strand silver drawn wire; the openwork of each petal is filled or partially filled with a substance which is perhaps degraded niello or enamel; incomplete copper-alloy back; an oval-section copper-alloy wire simple looped shank pierces the back and is soldered to the underside of the front; D *c*.19mm; shank L *c*.18mm. *South Somerset.*

Composite two-piece cast or sheet silver-gilt ground-supported filigree hanging-buttons with separate soldered drawn silver wire shanks

All are probably Norwegian *c*. late 16th-century, although this type reaches back to the 15th. Components are soldered together, as are a-d below. Hanging-buttons were made in a range of sizes (including of copper alloy) some of which are very large: those catalogued here are illustrated much enlarged. Hanging-buttons and dress-pin heads decorated with ground-supported filigree; or imitative ground-supported filigree, when separated from their shanks or stems respectively, are difficult to differentiate. Similarly, items which are perhaps beads may be heads from dress-pins or hanging-buttons with lost stems or pendent loops respectively. Seventeenth-century German or Dutch composite sheet-silver 'toggle' or 'link' buttons, which were attached to clothing by means of metal links, are really hanging-buttons. The latter have ground-supported filigree, granular, engraved, chased, repoussé or die-stamped decoration. No. 488 is possibly a dress-pin head. For comparison with ground-supported filigree buttons or hanging-buttons, shown below are four objects (a, b, c and d) with similar ornamentation.

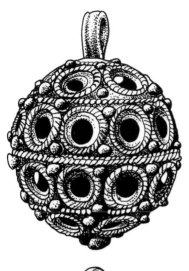

484. Biconvex; hollow; ground-supported filigree, granules and circular openwork; a flattened-section ribbed simple looped shank that perhaps extends through the sphere and is then riveted on the opposite side, thus forming an imitative pearl. *Provenance probably Norway.*

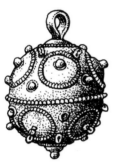

485. Biconvex; hollow; ground-supported filigree and granules; a flattened-section plain simple looped shank that perhaps extends through the sphere and is then riveted on the opposite side, thus forming an imitative pearl. *Provenance probably Norway.*

486. Biconvex; hollow; ground-supported filigree, granules and multiple multi-rayed sun openwork; a flattened-section ribbed simple looped shank is soldered to the back. *Provenance probably Norway.*

487. Biconvex; hollow; ground-supported filigree and granules; a circular-section simple looped shank that perhaps extends through the sphere and forms a simple loop on the opposite side; suspended from this loop is a separate looped imitative pearl pendant. *Provenance probably Norway.*

488. Biconvex; hollow; ground-supported filigree and granules; heavily gilded overall; (?) shank broken off; D 9mm. *South Somerset.*

a.   *c*. late 16th-century hollow-cast silver-gilt biconvex bead, possibly from a paternoster, decorated with ground-supported filigree and imitative pearl granulation; D 12mm. *South-East Dorset.*

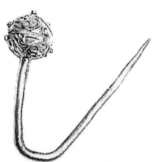

b.   *c*. late 16th-century hollow-cast silver-gilt biconvex-headed and drawn wire dress-pin decorated with ground-supported filigree and imitative pearl granulation; bent stem; D 14mm. *South-West Lincolnshire.*

c.   *c*. late 16th-century Read Class A, Type 2 hollow-sheet silver-gilt and drawn wire trefoil-shaped hat-hook decorated with ground-supported filigree and imitative pearl granulation and a sexfoil; body L 23.75mm; W 23.5mm. *South-East Dorset.* After Read 2008, no. 716.

d.   *c*.7th-century Anglo-Saxon sheet copper-alloy disc brooch decorated with soldered copper-alloy strips forming a voided saltire with black glass tiles (some missing or damaged) and gilt-copper-alloy ground-supported filigree; six drilled holes in the back-plate; D 28mm. *South-West Lincolnshire.* After Read 2001, no. 769.

Composite drawn silver wire button with separate drawn silver wire indeterminate double shank

In SCMT is a shoe-buckle and a silver button, both reputedly worn by the Duke of Monmouth during the Monmouth Rebellion in 1685. The buckle is of a type attributed to *c*.1660 – *c*.1720s, but by how much the button antedates the rebellion is unknown; it is perhaps even 16th century. Both the buckle and the button are enclosed in a small glass-fronted wooden box, which is probably of a somewhat later date, and impossible to remove. This restriction therefore precludes close inspection of the button; however, it appears to be crafted from single drawn strands of possibly twisted silver wire, although precisely how many pieces it comprises is uncertain. It possibly has an internal metal or organic spherical core, around which the wirework is formed, or, alternatively, such a core (usually beeswax) was melted out after manufacture. Strictly, this button is not filigree, for some wires overlap.

Fig 16. Contemporary painting of the Duke of Monmouth, *c*.1680s, oil on canvas. In SCMT.

123

Fig 17. Boxed shoe-buckle and silver button, reputedly worn by the Duke of Monmouth. *Somerset*. SCMT, acc. no. TTNCM 126/2002/2.

489. Biconvex; hollow with a possible core; drawn silver single strand (?) twisted wire crafted in alternate herring-bone and four-stranded with lateral bands segments; the shank is indeterminate but appears to be crossed twin simple loops, each of which is possibly formed from a silver drawn (?) twisted wire spiralled around a core; possible sporadic gilding overall; D *c.*14mm. *Somerset*. SCMT, acc. no. TTNCM 126/2002/2.

Composite three-piece part cast part die-stamped copper-alloy buttons with separate soldered drawn copper-alloy wire shanks

Nos 491-95 are from a *c.*17th-century context; no. 490 is probably of the same period. Components are soldered together. Unless otherwise stated, decoration is moulded-in-relief.

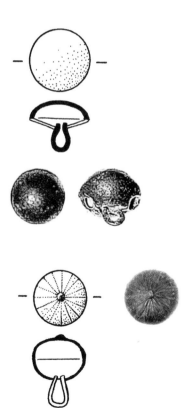

490. Biconvex; hollow; cast front; sheet back; undecorated; a circular-section simple looped shank; D 16mm; shank L 6.5mm. *South Somerset*.

491. Biconvex; hollow; cast front; sheet back; undecorated; the back has two blow-holes; a circular-section simple looped shank; (note anaerobic gilding). *River Thames foreshore, London*.

492. Biconvex; hollow; cast front; sheet back; multiple ridges radiating from a central pellet; the back has two blow-holes; sporadic white-metal coating overall; a circular-section simple looped shank; D 15mm; shank L 6mm. *River Thames foreshore, London*.

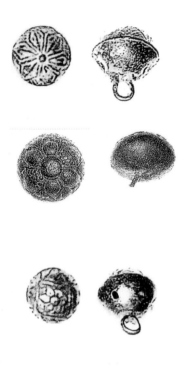

493. Biconvex; hollow; cast front; sheet back; a central pellet and a sexfoil; the back has two blow-holes; a circular-section simple looped shank; (note anaerobic gilding). *River Thames foreshore, London.*

494. Biconvex; hollow; cast front; sheet back; a central pellet and a sexfoil, two of the petals are in the form of small multifoils; incomplete, a circular-section simple looped shank broken off; (note anaerobic gilding). *River Thames foreshore, London.*

495. Biconvex; hollow; cast front; sheet back; a sexfoil and a central annulet within a voided lozenge on a hatched field; the back has two blow-holes; a circular-section simple looped shank; (note anaerobic gilding). *River Thames foreshore, London.*

Composite three-piece cast copper-alloy and sheet iron buttons with separate soldered drawn copper-alloy wire shanks
No. 496 came from a *c*.17th-century context; no. 497 is probably of the same period. Components are soldered together. Decoration on no. 496 is engraved and no. 497 moulded-in-relief and engraved.

496. Biconvex; hollow; cast copper-alloy front with a voided triangle hatched across each angle, a hatched field; sheet iron back; a circular-section simple looped shank. *River Thames foreshore, London.*

497. Discoidal; convex; an engraved voided eight-spoked wheel with a central pellet within a circle, hatched in the angles; white-metal coated; front of a biconvex hollow button; incomplete, back and either a separate soldered copper-alloy or iron wire shank or integral shank missing; rust inside and around the base; D 15mm. *South Somerset.*

Composite four-piece drawn wire or sheet copper-alloy buttons with separate soldered drawn copper-alloy wire shanks

Tentatively attributed as *c*.16th century, although unproven; perhaps as late as 18th – 19th century. Components are soldered together. Decoration on no. 498 is knurled and no. 499 is die-stamped. This type of button is also known made from sheet silver.

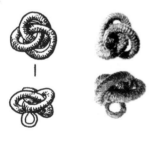

498. Sub-spherical; three interlinked split rings form a knot; multiple circumferential ridges; a circular-section simple looped shank; D 13mm; shank L 4mm. *North-West Kent. After TH, August 2004.*

499. Spherical; three interlinked split rings form a knot; each ring is rectangular-section with an upturned edge and has a band of foliate scrollwork on a pitted field; a circular-section simple looped shank; D 10mm; shank L *c*.5mm. *South Somerset.*

8: Post-Medieval Cuff Links

Members of the Society of Thames Mudlarks have recovered numerous metal cuff-links from stratified layers of the River Thames foreshore in London. Insofar as the capital goes, this datable evidence suggests that cuff-links first became fashionable around the beginning of the last quarter of the 17th century. Although reputed late 17th-century cuff-links are commonly found on inland sites, these seem to be restricted to a few well-known types. Learned opinion is divided as to whether glass-fronted cuff-links are earlier than mid-18th century. Single sections of cuff-links are frequently misidentified as buttons (especially types catalogued here as nos. 500-01). Unless otherwise stated, decoration is moulded-in-relief. All of the types catalogued here are known made in either precious- or base-metal.

Cast one-piece silver cuff-link with integral drilled shank
Cuff-links bearing this device are invariably associated with either the Restoration of King Charles II in 1660 or his marriage to Catherine of Braganza in 1662, attributions which the London evidence possibly refutes. Decoration is engraved.

500. Discoidal; flat; twin hearts surmounted with a crown; a downwards canted and lateral rim; a rectangular-section trapezoid shank; incomplete, one fastener and link missing; D 14mm; shank L 3.5mm. *South-West Lincolnshire.*

Cast one-piece copper-alloy cuff-links with integral drilled shanks
Both are *c.* late 17th century. No. 502 is from a late 17th-century context.

501. Discoidal; flat; an engraved five-petalled rose with sepals; a downwards canted rim, the internal and external edges of which are slightly expanded; a rectangular-section rounded shank; incomplete, one fastener and link missing; D 20mm; shank L *c.*7mm. *South Devon.* After Read 1995, no. 985, erroneously described as a button.

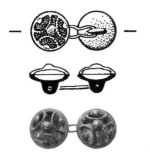

502. Each fastener is biconvex, hollow, with an openwork front comprising a central pellet and a circle within a voided and beaded curved-arm tribrach, a pellet borders each semi-circle, and a circular-section rounded shank; one fastener incomplete, a small section of tribrach arm broken off; a separate circular-section drawn wire link joins the two; D of both links 11mm; shank L 3.8mm; link L 13mm. *River Thames foreshore, London.* UKDFD 17254.

Composite three-piece sheet copper-alloy cuff-link with separate drawn wire shank
From a late 17th-century context. Decoration is die-stamped and engraved. Components are soldered together.

503. Each fastener is discoidal; hollow; with a convex front and flat back; and has a central pellet within a multi-spoked wheel and a voided lozenge, bordered with alternate arrows and lozenges; the back has two blow-holes and a D-section simple looped shank; a separate circular-section drawn wire link joins the two; D of both links 11mm; shank L 3.8mm; link L 11.5mm. *River Thames foreshore, London.* UKDFD 19929.

Cast one-piece lead/tin alloy cuff-links with integral undrilled shanks
Nos 506-7 are from a secure *c*.17th-century context; nos 504-5 are *c*. late 17th century. Decoration is moulded-in-relief and appliqués.

504. Discoidal; flat; a downwards canted rim and a vertical collet inset with a faceted green glass stone; incomplete, a ragged rim and a probable circular-section simple looped shank broken off, one fastener and indeterminate link missing; D *c*.13mm. *South-West Dorset.*

129

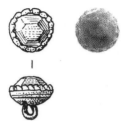

505. Plano-convex; a beaded rim with two circumferential grooves below; inset with a faceted amethyst-coloured glass stone; an oval-section simple looped shank; incomplete, one fastener and most of drawn iron wire link broken off; D12mm; shank L 3mm. *South-West Wiltshire.*

506. Each fastener is plano-convex, has a beaded rim with two circumferential linear grooves below; inset with a faceted clear glass stone and a circular-section simple looped shank; a separate cast lead/tin alloy circular-section hooked-link with openwork joins the two. *River Thames foreshore, London.*

507. Plano-convex; a beaded rim; a panel of foliate in whitish and black/brown champlevé enamel; a rounded-section rounded shank; incomplete, one fastener and link missing; D 12mm; shank L 4mm. *River Thames foreshore, London.*

Bibliography

Annable F K and Simpson D D A. *Guide Catalogue of the Neolithic and Bronze Age Collections in Devizes Museum*. 1964.

Ashelford, J. *A Visual History of Costume The Sixteenth Century*. 1983.

Bagley, P. *The Encyclopaedia of Jewellery Techniques*. 1986.

Bailey, G. *Detector Finds*. 1992.

— *Detector Finds 3*. 1997.

— *Buttons and Fasteners 500 BC-AD 1840*. 2004.

Biddle, M. *Object and Economy in Medieval Winchester*; Winchester Studies VII. 1990.

Biggs, N. *English Weights An Illustrated Survey*. 1992.

Biggs, N and Withers P. *Lead Weights The David Rogers Collection*. 2000.

Bishop, M C. The Camolmile Street Soldier Reconsidered. *London and Middlesex Archaeological Society Transactions 34*. 1983.

— *Finds from Roman Aldborough. A Catalogue of Small Finds from the Romano-British Town of Isurium Brigantum*. Oxbow Monograph 65. 1996.

Budge, E A W. *An Account of the Roman Antiquities Preserved in the Museum at Chesters, Northumberland*. London [1st Edn.]. 1903.

Clayton, M. *Catalogue of Rubbings of Brasses and Incised Slabs*. V&A Museum. 1968.

Cuddeford, M J. *Identifying Metallic Small Finds*. 1992 and 1994.

Cumming, V. *A Visual History of Costume The Seventeenth Century*. 1984.

East Anglian Archaeology. *Report No. 58; Norwich Survey*. 1993.

Edge and Paddock. *Arms and Armour of the Medieval Knight*. 1988.

Egan, G and Pritchard, F. *Dress Accessories c.1150 – c.1450 Medieval Finds From Excavations In London: 3*. 1991.

Epstein, D and Safro, M. *Buttons*. 1991.

Frere, S. 'Canterbury Excavations, Summer 1946' in *Archaeol Cantiana 67*.

Gaimster, D; Hayward, M; Mitchell, D; and Parker, K. Tudor Silver-Gilt Dress-Hooks: A New Class Of Treasure Find In England, in *The Antiquaries Journal, vol. 82*. 2002.

Geddes, J. 'The Small Finds', in Hare; J N (ed) 'Battle Abbey, The Eastern Range and the Excavations of 1978-80', *HBMCE Archaeol Rep 2*; London. 1985.

Geddes, J and Carter, A. '*Objects of Non-Ferrous Metal, Amber and Paste*', in Clarke and Carter. 1977.

Giltsoff, N and Robinson, P H. Buttons From Rural Wiltshire, in Wiltshire Archaeological and Natural History Society *Annual Bulletin No. 26*. 1980.

Goodall, A R and Goodall, I H. 'Copper-Alloy Objects', in Durham, B, 'Archaeological Investigations in St Aldates, Oxford', in *Oxoniensia 42*. 1977.

Hattatt, R. *Ancient Brooches and Other Artefacts*. 1989.

Hinton, D. *Medieval Jewellery*. 1982.

Houston, M G. *Medieval Costume in England and France the 13th, 14th and 15th Centuries.* 1965.

Hume, N H. *A Guide to Artifacts of Colonial America.* 1969.

Jowitt, R L P. *Salisbury.* 1951.

Kilbride-Jones, H E. *Celtic Craftsmanship in Bronze.* 1980.

Kite, E. *The Monumental Brasses of Wiltshire.* 1860.

Lightbown, R W. *Mediaeval European Jewellery.* 1992.

MacLeod, C. *Tudor and Jacobean Portraits in the National Portrait Gallery collection at Montacute House.* 1999.

Mackrell, A. *An Illustrated History of Fashion 500 Years of Fashion Illustration.* 1997.

Margeson, S. *Norwich Households: Post-Medieval Finds from Norwich Survey Excavations 1971–1978.* 1993.

Marshall, C. *Buckles Through The Ages.* 1986.

Meredith, A and G. *Buttons.* 2004.

Meredith, A and G, and Cuddeford, M J. *Identifying Buttons.* 1997.

Mills, N. *Roman Artefacts Found in Britain.* 1995.

— *Medieval Artefacts.* 1999.

— *Celtic and Roman Artefacts.* 2000.

Mitchener, M. *Medieval Pilgrim and Secular Badges.* 1986.

Murawski, G. *Benet's Artefacts of England and the United Kingdom.* 2000 and 2003.

Murdoch, T. *Treasures and Trinkets.* 1991.

Nevinson, J. 'Buttons and Buttonholes in the Fourteenth Century', *Costume, No. 11.* 1977.

Palmer, N. 'A Beaker Burial and Medieval Tenements in the Hamel, Oxford', in *Oxoniensia 45.* 1980.

Pearce, S M. *Bronze Age Metalwork in Southern Britain.* 1984.

Platt, C and Coleman-Smith, R. 'Excavations in Medieval Southampton 1953–69' 2, *The Finds*, Leicester. 1975.

PAS Finds Database.

Read, B A. *History Beneath Our Feet.* 1988 and 1995.

— *Metal Artefacts of Antiquity.* 2001.

— *Metal Buttons c.900 BC – c.AD 1700.* 2005.

— *Hooked-Clasps and Eyes.* 2008.

Rigold, S E. 'Eynsford Castle and its Excavations', in *Archaeol Cantiania 86.* 1971.

Rodríguez-Salgado, M J and the staff of the National Maritime Museum. *Armada 1588–1988.* 1988.

Saunders, P. (Ed). *Salisbury and South Wiltshire Museum Medieval Catalogue, Part 3.* 2001.

Savory, H N. *Guide Catalogue of the Iron Age Collections*. National Museum of Wales. 1976.

— *Guide Catalogue of the Bronze Age Collections*. National Museum of Wales. 1980.

Scott, M. *History of Dress: Late Gothic Europe 1400-1500*. 1980.

— *A Visual History of Costume The Fourteenth and Fifteenth Centuries*. 1986.

Smith, R A. *British Museum Guide to Early Iron Age Antiquities*. 1925.

Somerset, A. *Elizabeth I*. 1997.

Spencer, B. *Salisbury and South Wiltshire Museum Medieval Catalogue Part 2 Pilgrim Souvenirs and Secular Badges*. 1990.

Spencer, B. *Pilgrim Souvenirs and Secular Badges Medieval Finds From Excavations In London: 7*. 1998.

Spring, R. *Salisbury Cathedral*. 1987.

Stothard, C A. *The Monumental Effigies of Great Britain*. 1817.

Tebbutt, C F. 'St Neots Priory', in *Proc Cambridge Antiq Soc 59*. 1966.

Treasure Hunting. August and September 2004.

UKDF Database.

Untracht, O. *Jewellery Concepts and Technology*. 1982.

Waugh, N. *The Cut of Men's Clothes 1600–1900*. 1964.

Wild, J P. *Button-and-Loop Fasteners in the Roman Provinces*. 1970.